ez101 study keys

College Algebra

Second Edition

Lawrence S. Leff
Former Assistant Principal
Mathematics Supervisor
Franklin D. Roosevelt High School
Brooklyn, New York

All inquiries should be addressed to:
Barron's Educational Series, Inc.
250 Wireless Boulevard
Hauppauge, New York 11788
www.barronseduc.com

Library of Congress Catalog Card No. 2005049879

ISBN-13: 978-0-7641-2914-8
ISBN-10: 0-7641-2914-7

Library of Congress Cataloging-in-Publication Data
Leff, Lawrence S.
  College algebra / Lawrence S. Leff.—[2nd ed.].
    p. cm. — (Barron's EZ 101 study keys)
  ISBN-13: 978-0-7641-2914-8
  ISBN-10: 0-7641-2914-7
  1. Algebra.   I. Title. II. Series.

QA152.3.L44   2005
512.9—dc22   2005049879

PRINTED IN CANADA
9 8 7 6 5 4 3 2 1

CONTENTS

Theme 1 BASIC LAWS AND
OPERATIONS OF ALGEBRA

College Algebra reviews fundamental laws and operations of algebra while introducing more advanced algebraic concepts and methods. Theme 1 reviews some basic terminology and notation as well as some elementary algebraic operations.

Key 1 Sets, numbers, and variables

OVERVIEW *Algebra is generalized arithmetic in which letters, called **variables**, are used to represent numbers.*

Set: A set is a collection of objects, called **elements**, that are listed within braces, { }. If set S contains the first three lower-case letters of the alphabet, then $S = \{a, b, c\}$. Since the order of the elements within the braces does not matter, $\{b, a, c\} = \{a, b, c\}$.

Empty set: A set having no elements is the empty set. If S represents the set of elephants that can fly, then S is the empty set so $S = \{ \ \}$.

Null set: The empty set is sometimes called the *null set*. The null set is denoted by the symbol $\varnothing$.

Subsets: If each element of set B is also an element of set A, then set B is a subset of set A. If $B = \{2, 4\}$ and $A = \{1, \mathbf{2}, 3, \mathbf{4}, 5\}$, then set B is a subset of set A.

Set of natural (counting) numbers: The set $\{1, 2, 3, \ldots \}$ is the set of natural numbers. The trailing three dots inside the braces indicate that the pattern continues without ending.

Set of whole numbers: The set $\{0, 1, 2, 3, \ldots\}$ is the set of whole numbers.

Set of integers: The set $\{\ldots, -3, -2, -1, 0, 1, 2, 3, \ldots\}$ is the set of integers.

Set of rational numbers: The set of rational numbers is the set of all numbers that can be written as fractions in which the numerators are integers and the denominators are nonzero integers. Any integer can be put into this form by writing it as a fraction with a denominator of 1. For example, 3 can be written in fractional form as $\frac{3}{1}$. Thus, the sets of natural numbers, whole numbers, and integers are subsets of the set of rational numbers.

Rational numbers as decimal numbers: A rational number can be written as a nonending, repeating decimal number in which a set of one or more digits endlessly repeats, as illustrated below:

$$\frac{1}{4} = 0.25000\ldots, \frac{2}{3} = 0.66666\ldots, \text{ and } \frac{3}{11} = 0.272727\ldots$$

Also, every repeating decimal represents a rational number.

Set of irrational numbers: The set of irrational numbers contains all numbers that cannot be expressed as the quotient of two integers, such as $\sqrt{2}$ and π.

Irrational numbers as decimal numbers: Any nonending decimal number in which there is no repeating pattern of the same sequence of digits is irrational. Thus, $1.4142135\ldots (= \sqrt{2})$ and

0.13133133313333. . . are irrational numbers. Rational numbers have exact decimal equivalents, but irrational numbers do not.

Set of real numbers: The set of real numbers is the union of the set of rational numbers and the set of irrational numbers.

Real number line: The real number is a pictorial representation of the set of real numbers that resembles a "ruler-like" line on which an arbitrary point, called the **origin**, is labeled 0. Positive numbers lie in ascending order to the right of the origin, negative numbers appear in descending order to the left of the origin. As shown in Figure 1.1, each positive number and its negative counterpart (opposite) are located at points that are the same distance from the origin.

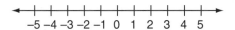

Figure 1.1 The Real Number Line

One-to-one correspondence: The set of real numbers and the set of points on a line can be paired so that each real number corresponds to exactly one point on the line and each point on the line corresponds to exactly one real number. The real number that is paired with a point is called the **coordinate** of that point.

Variable: Often represented by a letter from the alphabet such as x or y, a variable is a symbol that serves as a placeholder for an unknown member of a given set, called the **replacement set** or the **domain** of the variable. For example, suppose an unknown integer is increased by 8. We can represent this sum by calling the unknown number x and writing $x + 8$, where the set {. . ., –3, –2, –1, 0, 1, 2, 3, . . .} gives the possible replacements for x.

Constant. Unlike a variable, a constant has a fixed value. In the sum $x + 8$, 8 is a constant.

Generalizing in algebra: Mathematical laws and properties that apply to all the members of a set of numbers can be concisely stated using variables. For example, if a and b represent real numbers, then "$a + b = b + a$" states that the order in which *any* two real numbers are added does not matter.

Representing multiplication: The result of multiplying two or more quantities together is called a **product**. Each of the different quantities that are being multiplied to form a product is called a **factor** of the product. When writing products we try to avoid using the familiar arithmetic symbol for multiplication, x, since it may be confused with the variable x. The product of real numbers a and b may be represented in any of the following equivalent ways: by writing a and b consecutively, as in ab; or by placing a raised dot between a and b, as in $a \cdot b$; or by using parentheses, as in $(a)(b)$.

Key 2   Equality and inequality relations

OVERVIEW   *Special symbols are used to compare one real number to another. Unless otherwise indicated, letters used as variables always represent real numbers.*

Properties of equality: The following properties are assumed to be true:
- Reflexive property: $a = a$.
- Symmetric property: If $a = b$, then $b = a$.
- Transitive property: If $a = b$ and $b = c$, then $a = c$.
- Substitution property: If $a = b$, then in any statement in which a or b appears, one may replace the other.

Order on the number line: Real number a *is greater than* real number b, denoted by $a > b$, if and only if point a lies to the right of point b on the number line. As shown in Figure 1.1, –2 lies to the right of –5 on the number line so $-2 > -5$. Equivalently, –5 *is less than* –2 which is written as $-5 < -2$.

Transitive property of inequality: If $a < b$ and $b < c$, then $a < c$. For example, if $2 < 3$ and $3 < 4$, then $2 < 4$.

Trichotomy property: For any two real numbers exactly one of the following three statements is true:
1. $a < b$ (a is less than b)
2. $a = b$ (a is equal to b)
3. $a > b$ (a is greater than b)

The symbols ≠, ≥, and ≤ : If a and b are not equal, we can write $a \neq b$. We can indicate that a may be equal to *or* greater than b by writing $a \geq b$. The notation $a \leq b$ means a is less than *or* equal to b.

Combining inequality statements: If x is between a and b (with $a < b$) then $x > a$ and $x < b$. The two inequalities can be replaced by the single inequality $a < x < b$, which means that x is simultaneously greater than a and less than b, as shown in Figure 1.2.

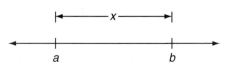

Figure 1.2 Combining Inequalities

The absolute value of a: This is denoted by $|a|$ and defined so that

$$|a| = a \text{ if } a \geq 0; \text{ and } |a| = -a \text{ if } a < 0$$

Thus, $|+4| = 4$ and $|-4| = -(-4) = 4$.

Key 3 Field properties of real numbers

OVERVIEW *In order to be able to work with real numbers, we need to make some basic assumptions about how they behave. These assumptions are called the **field properties** of real numbers.*

Binary operation: A binary operation, such as addition, works on exactly two numbers of a set at a time. If the outcome of the operation is also a member of the same set, then the set is said to be **closed** under this operation. The set of counting numbers is closed under addition, but not closed under subtraction since the difference between two counting numbers may not be a counting number, as in $2 - 3 = -1$.

Field properties of real numbers: The set of real numbers forms a field since we assume that each of the six field properties listed in Table 1.1 is true for real numbers a, b, and c. A subset of the real numbers is not necessarily a field. For example, the set of integers is not a field since it lacks field property F_5. Although each integer has an additive inverse, each integer does not have a multiplicative inverse. For example, the multiplicative inverse (reciprocal) of 2 is $\frac{1}{2}$ which is *not* a member of the set of integers.

Table 1.1   Field Properties for the Set of Real Numbers

Field Property	Addition	Multiplication
F_1: Closure	$a + b$ is a real number	$a \cdot b$ is a real number
F_2: Commutative	$a + b = b + a$	$ab = ba$
F_3: Associative	$(a + b) + c = a + (b + c)$	$(ab)c = a(bc)$
F_4: Identity	0 is the real number such that $a + 0 = 0 + a = 0$	1 is the real number such that $a \cdot 1 = 1 \cdot a = 0$
F_5: Inverse	For each real number a, $-a$ exists such that $a + (-a) = (-a) + a = 0$	For each real number a except 0, $\frac{1}{a}$ exists such that $a\left(\frac{1}{a}\right) = \left(\frac{1}{a}\right)a = 1$
F_6: Distributive	$a(b + c) = ab + ac$ and $(b + c)\,a = ba + ca$	

5

Definitions of subtraction and division: For any real numbers a and b,

- The **difference** $a - b$ is defined in terms of addition as $a + (-b)$. For example, $3 - 8 = (3) + (-8) = -5$.

- The **quotient** $a \div b$ is defined in terms of multiplication as $a\left(\dfrac{1}{b}\right)$, where a is called the **dividend**, b is called the **divisor**.

We assume $b \neq 0$. If dividing by 0 were allowed, then $3 \div 0$ would be equal to some number n. Since $3 \div 0 = n$ implies $3 = n \cdot 0$ and $n \cdot 0 = 0$, we obtain the contradictory statement, $3 = 0$. To avoid this type of contradiction, *division by zero is not defined.* Whenever a variable appears in a denominator of a fraction we assume that it cannot have a value that makes the denominator 0. Thus, in $\dfrac{3}{x - 2}$, x cannot be equal to 2.

Key 4 Interval notation

OVERVIEW *A continuous portion of the real number line corresponds to an **interval**. The numbers that belong to an interval can be indicated by writing the left and right boundary points of the interval as an ordered pair.*

Bounded intervals: If $a < b$, then the set of all real numbers from a to b forms a bounded interval with **endpoints** a and b. An **open interval** does not include its endpoints, and a **closed interval** includes both of its endpoints. A **half-open interval** (or half-closed interval) includes one endpoint, but not the other. The four possibilities are summarized in Table 1.2. On a graph, a darkened circle on an endpoint means the interval contains that endpoint. An open circle around an endpoint means that endpoint is not included.

Table 1.2   Intervals from a to b Where $a < b$

Type of Interval	Inequality Notation	Interval Notation	Graph of Interval
Open	$a < x < b$	(a, b)	
Half-Open	$a \le x < b$	$[a, b)$	
Half-Open	$a < x \le b$	$(a, b]$	
Closed	$a \le x \le b$	$[a, b]$	

Unbounded intervals: Intervals that extend without bound on one or both sides can be represented by the symbol for "infinity" (unbounded), ∞ . The symbol ∞ does not represent a real number. If a is a real number, then

- $(a, +\infty)$ means $x > a$.
- $[a, +\infty)$ means $x \ge a$.
- $(-\infty, a)$ means $x < a$.
- $(-\infty, a]$ means $x \le a$.

The entire real number line corresponds to the interval $(-\infty, +\infty)$.

Key 5   Integer exponents

OVERVIEW   *Repeated multiplication of the same quantity may be written in a more compact form using an exponent that tells the number of times the quantity appears as a factor in a product.*

x^n: This expression denotes the product in which x appears as a factor n times. Thus, $2^4 = 2 \cdot 2 \cdot 2 \cdot 2 = 16$. The symbol x^n is read as "x raised to the nth power" where x is called the **base** and n is the **exponent** of the base. When a variable appears without an exponent, the exponent is understood to be 1. Thus, $x = x^1$.

Laws of exponents: Powers of the same nonzero base can be multiplied by adding their exponents, or can be divided by subtracting their exponents. These rules are summarized in Table 1.3 where a and b are not 0, and x, y, and z are positive integers.

Table 1.3   Laws of Positive Integer Exponents

Law	Rule	Example
Multiplication	$a^x \cdot a^y = a^{x+y}$ $a^x \cdot b^x = (ab)^x$	$n^5 \cdot n^2 = n^7$ $m^3 \cdot n^3 = (mn)^3$
Division	$a^x \div a^y = a^{x-y}\,(x > y)$ $a^x \div a^y = \dfrac{1}{a^{y-x}}\,(y > x)$ $a^x \div b^x = \left(\dfrac{a}{b}\right)^x$	$n^5 \div n^2 = n^3$ $n^3 \div n^7 = \dfrac{1}{n^4}$ $m^3 \div n^3 = \left(\dfrac{m}{n}\right)^3$
Power of a Power	$(a^x)^y = a^{xy}$	$(n^5)^2 = n^{10}$
Power of a Product	$(a^x b^y)^z = a^{xz} b^{yz}$	$(mn^4)^3 = m^3 n^{12}$

Zero exponent: Any nonzero quantity raised to the zero power is 1. Thus, $x^0 = 1$ provided $x \neq 0$. The expression 0^0 is undefined.

Negative-integer exponents: If n is a positive integer and $x,\ y \neq 0$, then

$$x^{-n} = \frac{1}{x^n}, \quad \frac{1}{x^{-n}} = x^n$$

and

$$\left(\frac{x}{y}\right)^{-n} = \left(\frac{y}{x}\right)^n = \frac{y^n}{x^n}$$

Example: 1. $2^{-3} = \dfrac{1}{2^3} = \dfrac{1}{8}$

Example: 2. $\dfrac{1}{4^{-2}} = 4^2 = 16$

Scientific notation: Scientific notation is used to express a positive number as a number greater than or equal to 1 and less than 10 multiplied by an integer power of 10.

- To write a number greater than 10 in scientific notation:
  1. Move the decimal point k places to the left so that the number is now between 1 and 10.

     $1\,2\,0\,0\,0\,0\,0\,.$
     $k = 6$

  2. Multiply by 10^k.

     1.2×10^6

     Thus,

     $1{,}200{,}000 = 1.2 \times 10^6$

- To write a positive number less than 1 in scientific notation:
  1. Move the decimal point k places to the right so that the number is now between 1 and 10.

     $0\,.\,0\,0\,0\,0\,3\,0\,8$
     $k = 5$

  2. Multiply by 10^{-k}.

     3.08×10^{-5}

     Thus,

     $0.0000308 = 3.08 \times 10^{-5}$

- To write a positive number between 1 and 10 in scientific notation, multiply the number by 10^0. In scientific notation, $6.14 = 6.14 \times 10^0$.

Order of operations: Arithmetic operations are not necessarily performed in the order in which they are encountered. Instead:

1. First, evaluate powers.
2. Next, perform multiplications and divisions, working from left to right.
3. After that, perform additions and subtractions, again working from left to right.

If an expression contains parentheses or brackets, then the expression inside the grouping symbol is evaluated first.

Example 1.
$$5 + 18 \div (4 - 1)^2 = 5 + 18 \div 3^2$$
$$= 5 + 18 \div 9$$
$$= 5 + 2$$
$$= 7$$

Example 2. If $x = -2$ and $y = 5$, then
$$-y^2 + 3x^2 = -5^2 + 3(-2)^2$$
$$= -25 + 3(4)$$
$$= -25 + 12$$
$$= -13$$

Taking opposites: A negative sign that precedes a parenthesized expression may be interpreted as "take the opposite of whatever is inside the parentheses." Thus, $-(-5) = 5$. Sometimes it is helpful to replace the negative sign by -1 and to multiply, using the distributive property if necessary. For example,
$$-(2x - 3) = -1(2x - 3)$$
$$= (-1)2x + (-1)(-3)$$
$$= -2x + 3$$

Key 6 Polynomials

OVERVIEW *Variables and numbers may be multiplied and their products combined to produce expressions called **polynomials**. Since polynomials represent real numbers, they can be added, subtracted, multiplied, and divided using the properties of real numbers.*

Monomial: A monomial is a number, a variable, or the indicated *product* of a number and one or more variables with positive exponents; examples are 3, $-2xy$, and $5a^3b^2$. The numerical factor of a monomial is called its **coefficient**. The numerical coefficient of $-3ab$ is -3. The numerical coefficient of x^2 is 1 since $x^2 = 1 \cdot x^2$.

Like monomials: Monomials that have the same variables raised to the same powers, as in $3a^2b$, $-5a^2b$, and a^2b, are like monomials. They can be added or subtracted by combining their numerical coefficients.

Example 1. $3x^2 + 4x^2 = (3 + 4)x^2 = 7x^2$

Example 2. $7mn - mn = (7 - 1)mn = 6mn$

Polynomial: A polynomial is a monomial or the sum of two or more monomials. Each monomial in the sum is called a **term** of the polynomial. A polynomial having two unlike terms, like $2x + 3y$, is called a **binomial**. Since the polynomial $x^2 - 3x + 7$ has three unlike terms, it is called a **trinomial.**

The degree of a polynomial: The largest number obtained by adding the exponents of any term of a polynomial is its degree. The degree of $4xy + 7$ is 2 since $4xy = 4x^1y^1$ and $1 + 1 = 2$. The degree of $2x^3 - 5x^2 + 4x - 7$ is 3; and the degree of $x^2y + x^2y^3 + y^4$ is 5 since the sum of the exponents in the middle term is $2 + 3 = 5$. The degree of a polynomial is 0 when its only term is a nonzero constant.

Standard form: A polynomial in one variable is in standard form when the powers of the variable decrease in value as the terms are read from left to right. The polynomial $8x^3 + 6x^4 + 5 - 2x$ is not in standard form until its terms are rearranged so that the resulting polynomial is $6x^4 + 8x^3 - 2x + 5$.

Operations with polynomials:

- To add polynomials, collect and then add their like terms.

  Example:
  $$(4x^2 + 3y - 8) + (5x^2 - 2y + 7) = (4x^2 + 5x^2) + (3y - 2y) + (-8 + 7)$$
  $$= 9x^2 + y - 1$$

- To subtract polynomials, take the opposite of each term of the polynomial that is being subtracted, then add the two polynomials.

  Example:
  $$(3a^2 - 7a + 5) - (2a^2 - 4a - 1) = (3a^2 - 7a + 5) + (-2a^2 + 4a + 1)$$
  $$= (3a^2 - 2a^2) + (-7a + 4a) + (5 + 1)$$
  $$= a^2 - 3a + 6$$

- To multiply monomials, multiply their coefficients and then multiply the variable factors by *adding* the exponents of like variables.

  Example:
  $$(-2a^2b)(4a^3b^2) = (-2)(4)(a^2a^3)(bb^2)$$

  Note that $b = b^1$: $\quad = -8(a^{2+3})(b^{1+2})$
  $$= -8a^5b^3$$

- To divide monomials, divide their numerical coefficients and then divide the variable factors by *subtracting* the exponents of like variables.

  Example:
  $$\frac{14a^5b^4c}{21ab^3c^2} = \frac{2 \cdot 7}{3 \cdot 7}(a^{5-1})(b^{4-3})(c^{1-2})$$
  $$= \frac{2}{3}a^4b^1c^{-1} = \frac{2}{3}\frac{a^4b}{c}$$

- To multiply a polynomial by a monomial, multiply each term of the polynomial by the monomial, and then add the resulting products.

  Example:
  $$x^2(3x^2 - 5x + 1) = (x^2)(3x^2) + x^2(-5x) + x^2(1)$$
  $$= 3x^4 - 5x^3 + x^2$$

- To multiply a binomial by a polynomial, multiply each term of the polynomial by the binomial and then simplify using the distributive law.

Example:
$$(2x - 3)(x^2 - 5x + 4) = (2x - 3)\,x^2 + (2x - 3)\,(-5x) + (2x - 3)(4)$$
$$= (2x^3 - 3x^2) + (-10x^2 + 15x) + (8x - 12)$$
$$= 2x^3 - 13x^2 + 23x - 12$$

- To divide a polynomial by a monomial, divide each term of the polynomial by the monomial and then add the quotients.

Example:
$$\frac{12x^3 - 3x^2 + 6x}{3x} = \frac{12x^3}{3x} - \frac{3x^2}{3x} + \frac{6x}{3x}$$
$$= 4x^2 - x + 2$$

FOIL method for multiplying binomials: This method offers a short-cut for multiplying two binomials horizontally. The product of the binomials $(a + b)$ and $(c + d)$ is equal to the sum of the products of their **F**irst terms $(a \cdot c)$, **O**uter terms $(a \cdot d)$, **I**nner terms $(b \cdot c)$, and **L**ast terms $(b \cdot d)$.

Example:

$$
\begin{array}{cccccccc}
& & \mathbf{F} & & \mathbf{O} & & \mathbf{I} & & \mathbf{L} \\
(3x - 7)(2x + 5) & = & \overbrace{(3x)(2x)} & + & \overbrace{[(3x)(5)} & + & \overbrace{(-7)(2x)]} & + & \overbrace{(-7)(5)} \\
& = & 6x^2 & + & [15x & -14x] & & + & (-35) \\
& = & 6x^2 & + & & x & & & - 35
\end{array}
$$

Squaring binomials: The square of a binomial is a trinomial having one of the following forms:

- $(a + b)^2 = a^2 + 2ab + b^2$
- $(a - b)^2 = a^2 - 2ab + b^2$

Example:

To expand $(x + 2y)^2$, use the formula
$$(a + b)^2 = a^2 + 2ab + b^2$$
where $a = x$ and $b = 2y$. Thus,
$$(x + 2y)^2 = x^2 + 2(x)(2y) + (2y)^2$$
$$= x^2 + 4xy + 4y^2$$

Long division of polynomials: When the divisor of a polynomial is not a monomial, the quotient is obtained using a procedure similar to the one used to divide whole numbers in arithmetic. For example, proceed as follows to divide:

$$6x^3 - x^2 - 8x - 7 \text{ by } 2x - 3$$

Step 1. Divide the first term of the dividend by the first term of the divisor to obtain the first term of the quotient. The result of dividing $6x^3$ by $2x$ is $3x^2$.

$$2x - 3 \overline{\smash{\big)}\, 6x^3 - x^2 - 8x - 7} \quad \overset{3x^2}{}$$

Step 2. Multiply the entire divisor by the first term of the quotient and then subtract this product from the dividend. The difference (remainder) is $8x^2 - 8x$.

$$
\begin{array}{r}
3x^2 \\
2x - 3 \overline{\smash{\big)}\, 6x^3 - x^2 - 8x - 7} \\
\underline{^{-}\ 6x^3 - 9x^2 } \\
8x^2 - 8x
\end{array}
$$

Step 3. Using this remainder as the new dividend, repeat Steps 1 and 2 until a constant remainder is obtained.

$$
\begin{array}{r}
3x^2 + 4x + 2 \\
2x - 3 \overline{\smash{\big)}\, 6x^3 - x^2 - 8x - 7} \\
\underline{^{-}\ 6x^3 - 9x^2 } \\
8x^2 - 8x \\
\underline{^{-}\ 8x^2 - 12x } \\
4x - 7 \\
\underline{^{-}\ 4x - 6 } \\
-1
\end{array}
$$

The quotient is $3x^2 + 4x + 2$ with a remainder of -1.
Thus,

$$\frac{6x^3 - x^2 - 8x - 7}{2x - 3} = 3x^2 + 4x + 2 + \frac{-1}{2x - 3}$$

Step 4. Check by multiplying the quotient by the divisor and then adding any remainder. The check is left for you.

Key 7   Factoring polynomials

OVERVIEW   *Factoring a polynomial means writing the polynomial as the product of two or more lower degree polynomials each of which is called a **factor** of the original polynomial.*

Factoring over the set of integers: We usually factor a polynomial so that, if the original polynomial has integer coefficients, so does each of its factors.

Removing the greatest common monomial factor: If all the terms of a polynomial contain a common monomial factor, then it can be removed by using the reverse of the distributive property.

Example 1:   The greatest common factor (GCF) of the terms of $3x^4 + 12x$ is $3x$ since 3 and x are the greatest numerical and variable factors common to each term of the polynomial. Thus, $3x^4 + 12x = 3x (x^3 + 4)$. You can check that the factorization of $3x^4 + 12x$ is correct by multiplying the two factors together and verifying that the product is the original polynomial.

Example 2:   The greatest number that evenly divides the numerical coefficients in $24a^3b - 30a^2b^5$ is 6. The greatest powers of a and b that are common to both terms of the polynomial are a^2 and b^1. Hence, the GCF of the two terms of the polynomial is $6a^2b$. The corresponding factor of the polynomial can be obtained by dividing the polynomial by $6a^2b$ which gives $4a - 5b^4$. Thus, $24a^3b - 30a^2b^5 = 6a^2b(4a - 5b^4)$.

Factoring by grouping: Some polynomials can be factored by grouping their terms so that the reverse of the distributive property can be applied in successive stages. For example, the polynomial $xz + yz + xw + yw$ can be factored by first grouping the first two terms and the last two terms together, as in $z(x + y) + w(x + y)$, and then factoring out $(x + y)$. Thus,

$$xz + yz + xw + yw = z(x + y) + w(x + y)$$
$$= (x + y)(z + w)$$

Example:

$$8mx + 4px - 6m - 3p = (8mx + 4px) + (-6m - 3p)$$
$$= 4x(2m + p) - 3(2m + p)$$
$$= (2m + p)(4x - 3)$$

Special factoring patterns: Sometimes a polynomial can be factored by recognizing that it has the same form as a special polynomial whose factorization is already known. See Table 1.4.

Table 1.4 Special Factoring Patterns

Factorization Formula	Example
Difference of Two Squares: $a^2 - b^2 = (a - b)(a + b)$	Factor $4x^2 - 9$. Let $a = 2x$ and $b = 3$. Then, $4x^2 - 9 = (2x - 3)(2x + 3)$
Sum of Two Cubes: $a^3 + b^3 = (a + b)(a^2 - ab + b^2)$	Factor $x^3 + 8y^3$. Let $a = x$ and $b = 2y$. Then, $x^3 + 8y^3 = (x + 2y)(x^2 - 2xy + 4y^2)$
Difference of Two Cubes: $a^3 - b^3 = (a - b)(a^2 + ab + b^2)$	Factor $27x^3 - y^3$. Let $a = 3x$ and $b = y$. Then, $27x^3 - y^3 = (3x - y)(x^2 + 3xy + y^2)$

Factoring $x^2 + bx + c$: To factor a quadratic trinomial as the product of two binomials, use the reverse of FOIL.

Example:

$$x^2 + 7x + 10 = (x + \boxed{?})(x + \boxed{?}).$$

The missing terms in the binomial factors are the two numbers whose product is 10, the last term of $x^2 + 7x + \boxed{10}$, and whose sum is 7, the numerical coefficient of x in $x^2 + \boxed{7}x + 10$. Since $2 \times 5 = 10$ and $2 + 5 = 7$, the missing numbers are 2 and 5:

$$x^2 + 7x + 10 = (x + 2)(x + 5).$$

When factoring $x^2 + bx + c$:

- If c is positive, the constant terms in the binomial factors have the same sign, as in

$$y^2 - 7y + 12 = (y - 3)(y - 4) \quad \text{and} \quad y^2 + 7y + 12 = (y + 3)(y + 4).$$

- If c is negative, the constant terms in the binomial factors have opposite signs, as in

$$n^2 - 5n - 14 = (n + 2)(n - 7) \quad \text{and} \quad n^2 + 5n - 14 = (n - 2)(n + 7).$$

Always check that the factors work by multiplying the two binomial factors together and then comparing the product to the original quadratic trinomial.

Factoring $ax^2 + bx + c$ ($a > 1$): Factoring a quadratic trinomial becomes more difficult when the numerical coefficient of the x^2 – term is different from 1 because there are more possibilities to consider. To factor $3x^2 + 10x + 8$:

• Factor the x^2 – term and set up the binomial factors:

$$3x^2 + 10x + 8 = (3x + \boxed{?})(x + \boxed{?}).$$

• Determine the missing terms of the binomial factors. Find the two integers whose product is +8, the last term of $3x^2 + 10x + 8$, and that also make the sum of the outer and inner products of the terms of the binomial factors equal to $10x$. Thus, $3x^2 + 10x + 8 = (3x + 4)(x + 2)$.

Keep in mind that the placement of the factors of 8 matters. Although $(3x + 2)$ and $(x + 4)$ contain the correct factors of 8, the factors are not placed properly since the sum of the outer and inner products is $12x + 2x = 14x$, rather than $10x$.

Factoring completely: A polynomial is **factored completely** when each of its factors cannot be factored further. To factor completely, it may be necessary to use more than one factoring technique. Typically, we begin by removing a common monomial factor, if possible, and then factoring further, again if possible.

KEY EXAMPLE

Factor completely:

   (a) $3y^3 - 75y$ (b) $x^{4n} - 81$ (c) $a^3b + 2a^3 - b - 2$

Solution: (a) $3y^3 - 75y = 3y(y^2 - 25)$

Factor $y^2 - 25$: $= 3y(y - 5)(y + 5)$

   (b) $x^{4n} - 81 = (x^{2n} - 9)(x^{2n} + 9)$

Factor $x^{2n} - 9$: $= (x^n - 3)(x^n + 3)(x^{2n} + 9)$

   (c) $a^3b + 2a^3 - b - 2 = a^3(b + 2) - (b + 2)$

                   $= (b + 2)(a^3 - 1)$

Factor $a^3 - 1$: $= (b + 2)(a - 1)(a^2 + a + 1)$

Key 8 Radicals and rational exponents

OVERVIEW *The* k*th root of* x, *denoted by* $\sqrt[k]{x}$ *or* $x^{\frac{1}{k}}$, *is one of* k *identical real numbers whose product is* x.

Square root of *x*: A square root is one of two identical real numbers whose product is x. The square root of a nonnegative number x is denoted by $\sqrt{x}$. In $\sqrt{x}$, the symbol $\sqrt{}$ is called a **radical sign** and x is called the **radicand**. For example, $\sqrt{16} = 4$ since $4 \cdot 4 = 16$.

However, $\sqrt{-9}$ does not exist because the product of two identical real numbers can never equal a negative number.

Cube root: A cube root of a number is one of three identical real numbers whose product is the number. Thus, $\sqrt[3]{8} = 2$ since $2 \cdot 2 \cdot 2 = 8$ and $\sqrt[3]{-8} = -2$ since $(-2)(-2)(-2) = -8$. The *3* in $\sqrt[3]{-8}$ tells what root of x is to be taken and is called the **index** of the radical. In square root radicals the index is usually omitted. Thus, $\sqrt[2]{64}$ means $\sqrt{64}$.

The *k*th root of *x*: A real number b is the kth root of x, denoted by $\sqrt[k]{x}$, if and only if b satisfies the equation $b^k = x$. For example, 2 is a fourth root of 16 ($\sqrt[4]{16} = 2$) since $2^4 = 16$, and -4 is a cube root of -64 since $(-4)^3 = -64$. If $x < 0$ and k is even, then $\sqrt[k]{x}$ is not a real number so the kth root of x does not exist.

Principal *k*th root: The notation $\sqrt[k]{x}$ always refers to the principal kth root of x, which is the real valued root, if it exists, that has the same sign as x whenever $x \neq 0$. For example, although the two square roots of 16 are 4 and -4, only 4 is the *principal* square root of 16 since 4 has the same sign as 16. Thus, $\sqrt{16} = 4$.

Definition of $x^{\frac{1}{k}}$: The notation $x^{\frac{1}{k}}$ denotes the principal kth root of x provided that $x \geq 0$ whenever k is even. Thus,

$$x^{\frac{1}{k}} = \sqrt[k]{x}$$

Examples:

$$36^{\frac{1}{2}} = \sqrt{36} = 6 \text{ and } (-125)^{\frac{1}{3}} = \sqrt[3]{-125} = -5$$

Definition of $x^{n/k}$:   If $\sqrt[k]{x^n}$ is a real number, then

$$x^{n/k} = (\sqrt[k]{x})^n = \sqrt[k]{x^n} ,$$

where x is any real number, n is an integer, and k is a positive integer.

In evaluating expressions of the form $x^{\frac{n}{k}}$, it is usually easier to first take the kth root of x and then raise the result to the nth power, as in

$$27^{\frac{2}{3}} = (\sqrt[3]{27})^2 = (3)^2 = 9 .$$

Laws for rational exponents: The laws of positive-integer exponents given in Table 1.3 hold also for exponents that are rational numbers.

Key 9 Operations with radicals

OVERVIEW *Radicals may be multiplied and divided, provided that they have the same index. Radicals that have the same index and the same radicand can be added or subtracted.*

Multiplication rule for radicals: $c\sqrt[k]{a} \cdot d\sqrt[k]{b} = cd\sqrt[k]{ab}$, provided $ab \geq 0$ if k is even.

Example: $2\sqrt{5} \cdot 3\sqrt{7} = 6\sqrt{35}$

Quotient rule for radicals:

$$\frac{c\sqrt[k]{a}}{d\sqrt[k]{b}} = \frac{c}{d} \sqrt[k]{\frac{a}{b}}$$

(when $b \neq 0$ and $ab \geq 0$ if k is even).

Example: $\dfrac{12\sqrt{21}}{6\sqrt{7}} = \dfrac{12}{6} \sqrt{\dfrac{21}{7}} = 2\sqrt{3}$

Perfect square: Any term that can be expressed as the product of two identical factors is a perfect square. Forty-nine is a perfect square since $49 = 7 \cdot 7$.

Simplifying radicals: To simplify $\sqrt{N}$, write $\sqrt{N} = \sqrt{a} \cdot \sqrt{b}$ where a is the greatest perfect square factor of N. Then evaluate $\sqrt{a}$.

Example 1. $\sqrt{28}$ $=$ $\sqrt{4} \cdot \sqrt{7}$ $=$ $2\sqrt{7}$
Example 2. $\sqrt{48x^5}$ $=$ $\sqrt{16x^4} \cdot \sqrt{3x}$ $=$ $4x^2\sqrt{3x}$

In Example 2, x^4 is the greatest perfect square factor of x^5 since its exponent is the greatest power contained in x^5 that is divisible by 2, the index of the radical. Radicals with indexes other than 2 can be simplified using a similar procedure.

Example 3. $\sqrt[3]{24y^7}$ $=$ $\sqrt[3]{8y^6} \cdot \sqrt[3]{3y}$ $=$ $2y^2\sqrt[3]{3y}$

In Example 3, 8 is the greatest perfect cube factor of 24 and y^6 is the greatest perfect cube factor of y^7. The factor y^6 is obtained by finding the greatest factor of y^7 whose exponent is evenly divisible by the index of the radical, 3.

Like radicals: Like radicals have the same index and the same radicand. Like radicals can be added or subtracted by combining their coefficients, as in $2\sqrt{5} + 7\sqrt{5} = (2+7)\sqrt{5} = 9\sqrt{5}$. Sometimes unlike radicals can be combined after the radicals have been simplified.

Example:

$$3\sqrt{48} + \sqrt{75} = 3\sqrt{16}\sqrt{3} + \sqrt{25}\sqrt{3}$$
$$= 3 \cdot 4\sqrt{3} + 5\sqrt{3}$$
$$= 12\sqrt{3} + 5\sqrt{3}$$
$$= 17\sqrt{3}$$

Multiplying radical expressions: Two radical expressions that each have the form of a binomial can be multiplied using FOIL.

Example:

$$\overset{\text{F}\quad\text{O}\quad\text{I}\quad\text{L}}{(2+3\sqrt{5})(1-\sqrt{7}) = 2 - 2\sqrt{7} + 3\sqrt{5} - 3\sqrt{35}}$$

Conjugate radical expressions: Pairs of radical expressions of the form $(A + B\sqrt{C})$ and $(A - B\sqrt{C})$ are called conjugate radicals. The product of conjugate square root radicals does not include a radical since, using FOIL,

$$(A + B\sqrt{C})(A - B\sqrt{C}) = A^2 - B^2 C$$

Example: The product $(7 + 2\sqrt{3})(7 - 2\sqrt{3})$ can be obtained by letting $A = 7$, $B = 2$, and $C = 3$. Thus,

$$(7 + 2\sqrt{3})(7 - 2\sqrt{3}) = 7^2 - 2^2(3) = 49 - 12 = 37.$$

Rationalizing a denominator: This is the process of changing a fraction with a radical in its denominator into an equivalent fraction whose denominator has no radical. To rationalize a denominator of the form $B\sqrt{C}$, multiply the fraction by 1 in the form of $\dfrac{\sqrt{C}}{\sqrt{C}}$.

Example:

$$\frac{5}{\sqrt{3}} = \frac{5}{\sqrt{3}} \cdot \left(\frac{\sqrt{3}}{\sqrt{3}}\right) = \frac{5\sqrt{3}}{\sqrt{9}} = \frac{5\sqrt{3}}{3}$$

Similarly, to rationalize a denominator of the form $A + B\sqrt{C}$, multiply the numerator and the denominator of the fraction by the conjugate, $A - B\sqrt{C}$, of the denominator.

Example: Rationalize the denominator of $\dfrac{\sqrt{8}}{5-3\sqrt{2}}$ by using the conjugate of the denominator, $5+3\sqrt{2}$, as the rationalizing factor. Thus,

$$\frac{\sqrt{8}}{5-3\sqrt{2}} \cdot \left(\frac{5+3\sqrt{2}}{5+3\sqrt{2}}\right) = \frac{\sqrt{8}(5+3\sqrt{2})}{5^2 - 3^2(2)}$$

$$= \frac{5\sqrt{8} + 3\sqrt{16}}{25-18}$$

$$= \frac{5\sqrt{8} + 12}{7}$$

Theme 2 EQUATIONS AND INEQUALITIES

*F*inding solutions to various types of equations and inequalities is an important aspect of algebra. Solving a linear equation requires isolating the variable on one side of the equation and placing the constant terms on the other side. To do this we need to undo whatever operations are "attached" to the variable by performing their inverses. For example, if $x + 3 = 7$, we can cancel the effect of *adding* 3 to x by *subtracting* 3 from each side of the equation. This gives $x = 4$. Solving equations that are not linear may also involve performing inverse operations. If $\sqrt{x} = 5$, we solve for x by undoing the square root of x by raising both sides of the equation to the second power. This gives $x = 25$. To solve $x^2 - 3x = 0$, we undo the multiplication that led to $x^2 - 3x$, by factoring the left side of the equation as $x(x - 3) = 0$ and then setting each factor equal to 0 so that $x = 0$ or $x = 3$.

INDIVIDUAL KEYS IN THIS THEME

10	Linear equations
11	Linear inequalities
12	Absolute value equations and inequalities
13	Quadratic equations
14	Quadratic inequalities
15	Radical equations

Key 10 Linear equations

OVERVIEW *An **equation** states that two expressions represent the same number. An equation that contains a variable cannot be judged true or false until the variable is replaced with a number.*

Solving an equation: Solving an equation for a variable is the process of finding all values in the domain of the variable for which the equation is true. Each of the values is a **solution** or **root** of the equation. The collection of all roots of an equation is the **solution set**.

Conditional equation: An equation whose solution set may include some but not all of the values in the replacement set is a conditional equation.

Identity: An equation that is true for all possible replacements for the variable is an identity. The equation $2x + 1 = 2$ is a conditional equation since it is true only when x is replaced by $\frac{1}{2}$. The equation $3(x + 1) - 1 = 3x + 2$ is true for all possible replacements for x, so it is an identity.

Equivalent equations: These equations have the same solution set. The equations $x + 1 = 4$, $2x = 6$, and $x = 3$ are equivalent equations since they have the same solution, $x = 3$.

Properties of equations: The following properties may be needed when solving equations:

*   **P$_1$** Adding or subtracting the same quantity on both sides of the equation produces an equivalent equation. For example, if

$$x + 1 = 3,$$

then

$$x + 1 - (1) = 3 - (1)$$

so

$$x = 2.$$

*   **P$_2$** Multiplying or dividing both sides of the equation by the same *nonzero* quantity produces an equivalent equation. For example, if

$$3x = 6,$$

then

$$\frac{3x}{3} = \frac{6}{3}$$

so

$$x = 2.$$

Linear (first-degree) equation: In a linear equation the exponent of any variable is 1. For example, $2x - 1 = x + 3$ is a linear equation, but the equation $x^2 = 4$ is not linear. A linear equation in one variable, say x, is an equation that can be put into the form $ax + b = 0$, where a and b are constants with $a \neq 0$.

Strategy for solving linear equations: Use the properties of equations to obtain successively simpler equivalent equations until the variable is isolated by itself on one side of the equation and a constant term is alone on the other side.

KEY EXAMPLE

Solve for x: $\qquad\qquad 5x - 8 = 2x + 13$

Solution: Collect variable terms on the left side of the equation and the constant terms on the right side of the equation.

By $\mathbf{P_1}$:	$5x - 8 + (8) = 2x + 13 + (8)$
Simplify:	$5x = 2x + 21$
By $\mathbf{P_1}$:	$5x - (2x) = 2x + 21 - (2x)$
Simplify:	$3x = 21$
By $\mathbf{P_2}$:	$\dfrac{3x}{3} = \dfrac{21}{3}$
Simplify:	$x = 7$

KEY EXAMPLE

Solve for x: $\qquad\qquad 2(7 - 3x) = 3(x + 1) + 29$

Solution: Remove the parentheses by multiplying each term inside the parentheses by the number in front of the parentheses. Then collect variable terms on the left side of the equation and constants on the right side of the equation.

$$2(7 - 3x) = 3(x + 1) + 29$$

Use the Distributive Property:	$2(7) - 2(3x) = 3(x) + 3(1) + 29$
Simplify:	$14 - 6x = 3x + 32$
By $\mathbf{P_1}$:	$14 - 6x - (14) = 3x + 32 - (14)$
Simplify:	$-6x = 3x + 18$

By $\mathbf{P_1}$:	$-6x - (3x) = 3x + 18 - (3x)$
Simplify:	$-9x = 18$
By $\mathbf{P_2}$:	$\dfrac{-9x}{-9} = \dfrac{18}{-9}$
	$x = -2$

Checking roots: To check that a number is a root or solution of an equation, replace the variable by the root in the original equation and verify that the left-hand and the right-hand sides of the original equation represent the same number.

Example: To check that $x = 7$ is a root of $5x - 8 = 2x + 13$, replace x with 7 and then compare the values of the left-hand and the right-hand sides of the equation:

$$5(7) - 8 \overset{?}{=} 2(7) + 13$$

$$35 - 8 \overset{?}{=} 14 + 13$$

$$\checkmark$$

$$27 = 27$$

To solve linear equations with fractions: Clear the equation of its fractions by multiplying each term of the equation by the least common multiple (LCM) of the denominators of the fractional terms.

Example: To solve $\dfrac{x+1}{4} - \dfrac{2}{3} = \dfrac{x-7}{12}$,

write an equivalent equation without fractions by multiplying each term of the equation by 12, which is the smallest number into which 4, 3, and 12 divide evenly:

$$12\left(\frac{x+1}{4}\right) - 12\left(\frac{2}{3}\right) = 12\left(\frac{x-7}{12}\right).$$

The resulting equation is

$$3(x + 1) - 8 = x - 7,$$

whose solution, as you can verify, is $x = -1$.

Proportion: A proportion is an equation having the form $\dfrac{a}{b} = \dfrac{c}{d}$.

Cross-products rule: If

$$\frac{a}{b} = \frac{c}{d}, \text{ then } bc = ad.$$

Example: To solve the proportion

$$\frac{x+5}{4} = \frac{x-2}{3}$$

write the equivalent equation
$$4(x - 2) = 3(x + 5)$$
whose solution, as you can verify, is $x = 23$.

Literal equation: An equation with more than one letter or variable is a literal equation. Solving a literal equation for a specified letter or variable requires isolating this letter or variable in much the same way that an equation having a single variable is solved.

Example: To solve
$$ay = by + c \text{ for } y,$$

isolate y on the left side of the equation by subtracting by from both sides of the equation:

$$ay - by = by + c - by$$

Factor out y: $\quad y(a - b) = c$

Divide by the coefficient of y: $\quad \dfrac{y(a-b)}{a-b} = \dfrac{c}{a-b}$

$$y = \frac{c}{a-b} \text{ (provided } a \neq b)$$

Key 11 Linear inequalities

OVERVIEW *Replacing the equal sign in a linear equation with an inequality symbol creates a **linear inequality**. Linear inequalities are solved in much the same way that linear equations are solved.*

Properties of inequality: An inequality can be changed into an equivalent inequality by performing the following operations:

P₃ Add or subtract the same quantity on both sides of an inequality. For example, if $x - 1 < 3$, then adding 1 to each side gives the equivalent inequality, $x < 4$.

P₄ Multiply or divide both sides of an inequality by the same *positive* quantity. For example, if $3x < 6$, then dividing each side by 3 gives the equivalent inequality, $x < 2$.

P₅ Multiply or divide both sides of an inequality by the same *negative* quantity and then reverse the direction of the inequality sign. For example, if $-2x > 8$, then dividing each side by -2 and reversing the direction of the inequality from $>$ to $<$ gives the equivalent inequality, $x < -4$.

KEY EXAMPLE

Solve and graph the solution set of

$$1 - 2x < x + 13.$$

Solution: Isolate the variable on the left side of the inequality:

By **P₃**:
$$1 - 2x - 1 < x + 13 - 1$$
$$-2x < x + 12$$

By **P₃**:
$$-x - 2x < x + 12 - x$$
$$-3x < 12$$

By **P₅**:
$$\frac{-3x}{-3} > \frac{12}{-3} \quad \text{(direction of inequality is reversed)}$$
$$x > -4$$

The solution is $x > -4$, which when graphed consists of all points greater than, but not equal to, -4.

Compound inequality: The solution of the compound inequality $3 < 2x - 1 < 5$ must satisfy the two inequalities $3 < 2x - 1$ and $2x - 1 < 5$ simultaneously. Performing the same arithmetic operation with each member of a compound inequality produces an equivalent inequality. Thus, adding 1 to each member of $3 < 2x - 1 < 5$ produces the equivalent inequality $4 < 2x < 6$. Then dividing each member of this inequality by 2 gives $2 < x < 3$.

KEY EXAMPLE

Solve and graph the solution:

$$-1 \le \frac{4 - 7x}{3} < 6$$

Solution: Clear the fraction by multiplying each member of the inequality by 3:

$$(3)(-1) \quad \le \quad (3)\frac{4 - 7x}{3} \quad < \quad (3)(6)$$

$$-3 \quad \le \quad 4 - 7x \quad < \quad 18$$

$$-3 - 4 \quad \le \quad 4 - 7x - 4 \quad < \quad 18 - 4$$

$$-7 \quad \le \quad -7x \quad < \quad 14$$

Divide by -7 and reverse the direction of the inequality:

$$\frac{-7}{-7} \quad \ge \quad \frac{-7x}{-7} \quad > \quad \frac{14}{-7}$$

$$1 \quad \ge \quad x \quad > \quad -2$$

$$-2 \quad < \quad x \quad \le \quad 1$$

or, equivalently,

The solution, $-2 < x \le 1$, represents the interval $(-2, 1)$, the graph of which is shown.

Key 12 Absolute value equations and inequalities

OVERVIEW *Absolute value equations and inequalities are solved by writing equivalent statements that do not involve absolute value. This is accomplished using the following properties, where* c *is a positive number:*

- *(i) If* $|x| = c$, *then* $x = c$ *or* $x = -c$.
- *(ii) If* $|x| < c$, *then* $-c < x < c$.
- *(iii) If* $|x| > c$, *then* $x < -c$ *or* $x > c$.

To solve $|ax + b| = d$: After replacing x with $ax + b$ in property (i), write and then solve each equation,

$$ax + b = d \text{ or } ax + b = -d.$$

Example: If $|2x + 3| = 1$, then

$$2x + 3 = 1 \quad \text{or} \quad 2x + 3 = -1$$

$$2x = -2 \qquad\qquad 2x = -4$$

$$\frac{2x}{2} = \frac{-2}{2} \qquad\qquad \frac{2x}{2} = \frac{-4}{2}$$

$$x = -1 \quad \text{or} \qquad x = -2$$

KEY EXAMPLE

Solve and check: $|x + 3| = 2x$

Solution: If $|x + 3| = 2x$, then

$$x + 3 = 2x \quad \text{or} \quad x + 3 = -2x$$

$$-x + 3 = 0 \quad \text{or} \quad 3x + 3 = 0$$

$$-x = -3 \quad \text{or} \quad 3x = -3$$

$$x = 3 \quad \text{or} \quad x = -1$$

Check:

$$\text{Let } x = 3: \quad \text{or} \quad \text{Let } x = -1:$$
$$|x + 3| = 2x \quad \text{or} \quad |x + 3| = 2x$$
$$|3 + 3| = 2(3) \quad \text{or} \quad |-1 + 3| = 2(-1)$$
$$6 = 6 \quad \text{or} \quad 2 \neq -2$$

Reject this solution.

Hence, the only root of the equation is $x = 3$.

KEY EXAMPLE

Solve and check: $|2x + 1| = |2x - 9|$

Solution: In general, if $|A| = |B|$, then $A = |B|$ or $A = -|B|$, which means that $A = B$ or $A = -B$. Hence, if
$$|2x + 1| = |2x - 9|,$$
then

$$2x + 1 = 2x - 9 \quad \text{or} \quad 2x + 1 = -(2x - 9)$$
$$1 = -9 \qquad\qquad 2x + 1 = -2x + 9$$

Impossible result implies
the equation has no solution.

$$2x = -2x + 8$$
$$4x = 8$$
$$x = 2$$

The check is left for you.

To solve $|ax + b| < d$: After replacing x with $ax + b$ in property (ii), write and then solve the compound inequality

$$-d < ax + b < d.$$

If

$$|ax + b| \leq d,$$

then

$$-d \leq ax + b \leq d.$$

To solve $|ax + b| > d$: After replacing x with $ax + b$ in property (iii), write and then solve the compound inequality
$$ax + b < -d \text{ or } ax + b > d.$$

If

$$|ax + b| \geq d,$$

then

$$ax + b \leq -d \text{ or } ax + b \geq d.$$

KEY EXAMPLE

Solve for x:
$$\left|\frac{5x-7}{3}\right| < 1$$

Solution: If $\left|\frac{5x-7}{3}\right| < 1$, then $-1 < \frac{5x-7}{3} < 1$.

Isolate the variable, making sure to perform the same operations on both sides of the inequality:

$$(3)(-1) < 3\left(\frac{5x-7}{3}\right) < (3)(1)$$

$$-3+7 < 5x-7+7 < 3+7$$

$$4 < 5x < 10$$

$$0.8 < x < 2$$

The solution is the open interval (0.8, 2).

KEY EXAMPLE

Solve for x: $\qquad 11 < |4 - 3x|$.

Solution : If $\qquad 11 < |4 - 3x|$, then $\qquad |4 - 3x| > 11$, so

$$4 - 3x < -11 \qquad \text{or} \qquad 4 - 3x > 11$$

$$-3x < -11 - 4 \qquad \text{or} \qquad 4 - 3x - 4 > 11 - 4$$

$$-3x < -15 \qquad \text{or} \qquad -3x > 7$$

$$x > 5 \qquad \text{or} \qquad x < -\frac{7}{3}$$

The solution is $x > 5$ or $x < -\frac{7}{3}$.

Key 13   Quadratic equations

OVERVIEW   *A **quadratic equation** in* x *is an equation that can be written in the standard form* $ax^2 + bx + c = 0$ *(a ≠ 0). Some quadratic equations can be solved by factoring the quadratic polynomial.*

Zero product rule: If $a \cdot b = 0$, then either $a = 0$, $b = 0$, or both $a = 0$ and $b = 0$.

Solving quadratic equations by factoring: If the left side of the quadratic equation $ax^2 + bx + c = 0$ $(a \neq 0)$ can be factored, then by the zero product rule each linear factor may be equal to 0. The solutions to these linear equations are the roots of the quadratic equation. For example, if $x^2 + 4x = 0$, then $x(x + 4) = 0$. Since either factor can equal 0, $x = 0$ or $x + 4 = 0$. If $x + 4 = 0$, then $x = -4$. Hence, the two roots of the equation $x^2 + 4x = 0$ are $x = 0$ and $x = -4$.

KEY EXAMPLE

Solve and check: $3x^2 + 6x = 45$

Solution: Write the equation in the standard form $ax^2 + bx + c$:

$$3x^2 + 6x - 45 = 0.$$

To simplify the factoring, divide through by any common numerical factors:

$$\frac{3x^2}{3} + \frac{6x}{3} - \frac{45}{3} = \frac{0}{3}$$

$$x^2 + 2x - 15 = 0$$

Factor: $(x - 3)(x + 5) = 0$

Apply the zero product rule: $x - 3 = 0$ or $x + 5 = 0$

$x = 3$ or $x = -5$

The check is left for you.

KEY EXAMPLE

Solve for y and check: $(2y - 1)(y + 3) = 4$

Solution: Since the right side of the equation is not 0, the zero product rule cannot be used. Put the equation into standard form by multiplying the factors and then collecting all terms on the left side of the equation with 0 on the right side. The result is

$$2y^2 + 5y - 7 = 0.$$

Factoring and using the zero product rule leads to

$$2y + 7 = 0 \text{ or } y - 1 = 0.$$

Hence, the roots of the equation are $y = -\dfrac{7}{2}$ and $y = 1$.

The check is left for you.

Solving higher degree equations by factoring: Sometimes an equation of degree higher than 2 can be solved by factoring completely. For example, if

$$-n^3 - 5n^2 + 6n = 0,$$

then, after multiplying each term by -1, n can be factored out so

$$n(n^2 + 5n - 6) = 0.$$

Factoring the quadratic polynomial gives

$$n(n + 6)(n - 1) = 0.$$

If the product of three expressions is 0, then at least one of these expressions is 0. One root is $n = 0$. If $n + 6 = 0$, then $n = -6$. If $n - 1 = 0$, then $n = 1$. The three roots of the original third-degree equation are $n = 0$, and $n = -6$, and $n = 1$.

Solving equations having a quadratic form: If a nonlinear equation can be put into the form $au^2 + bu + c = 0$ where $a \neq 0$ and u is some variable expression, then the equation has a quadratic form. For example, since the equation $x^4 - 5x^2 + 4 = 0$ can be written as $(x^2)^2 - 5x^2 + 4 = 0$, it may be made to look like a familiar type of quadratic equation by substituting u for x^2. The transformed equation is $u^2 - 5u + 4 = 0$. Its roots are $u = 1$ and $u = 4$. If $u = x^2 = 1$, then $x = \pm 1$. If $u = x^2 = 4$, then $x = \pm 2$. The *four* roots of the original *fourth*-degree equation are $u = -1$, $u = +1$, $u = -2$, and $u = +2$.

KEY EXAMPLE

Solve and check: $\qquad x + \sqrt{x} - 6 = 0$

Solution: Since

$$x = (\sqrt{x})^2,$$

the given equation can be written as

$$(\sqrt{x})^2 + \sqrt{x} - 6 = 0.$$

Let $u = \sqrt{x}$; then the original equation becomes

$$u^2 + u - 6 = 0$$

whose roots are $u = 2$ and $u = -3$. If

$$u = \sqrt{x} = 2,$$

then $\qquad\qquad\qquad\qquad x = 4.$

If $\qquad\qquad\qquad\qquad u = \sqrt{x} = -3,$

then this solution must be rejected since $\sqrt{x}$ must be nonnegative. The original equation has the single root $x = 4$. The check is left for you.

You could, of course, have solved the equation directly by factoring. If $x + \sqrt{x} - 6 = 0$, then

$$(\sqrt{x} + 3)(\sqrt{x} - 2) = 0, \text{ so } \sqrt{x} = -3 \text{ or } \sqrt{x} = 2.$$

Reject $\sqrt{x} = -3$. Since $\sqrt{x} = 2$, $x = 4$.

The original equation could also have been solved by isolating the radical. Because $\sqrt{x} = 6 - x$,

$$x = (6 - x)^2 = 36 - 12x + x^2.$$

Rearranging terms gives $x^2 - 13x + 36 = 0$. Thus:

$$(x - 4)(x - 9) = 9, \text{ so } x = 4 \text{ or } x = 9.$$

You should verify that works in the original equation. Since $x = 9$ fails, it is rejected.

Key 14 Quadratic inequalities

OVERVIEW *Replacing the equal sign in a quadratic equation with an inequality symbol produces a **quadratic inequality**. The real roots of the related quadratic equation determine the endpoints of the possible solution intervals of the quadratic inequality.*

To solve a quadratic inequality by factoring: Write the quadratic inequality so that 0 is alone on the right side of the inequality. Solve the related quadratic equation. The roots of this equation, called **critical numbers**, divide the real number line into nonoverlapping **solution intervals**. If a test value from an interval satisfies the original inequality, then the interval from which it was selected belongs to the solution set.

KEY EXAMPLE

Solve for x: $\qquad\qquad x^2 - 2x < 3$

Solution: The original inequality in standard form is

$$x^2 - 2x - 3 < 0.$$

Factoring gives

$$(x + 1)(x - 3) < 0.$$

The roots of

$$(x + 1)(x - 3) = 0$$

are $x = -1$ and $x = 3$. These critical numbers divide the real number line into three intervals: $x < -1$, $-1 < x < 3$, and $x > 3$. Select a test value from each interval and determine whether it makes the inequality $(x + 1)(x - 3) < 0$ true or false. It may help to organize your work as follows:

Interval	Test Value	$(x + 1)(x - 3)$	$(x + 1)(x - 3) < 0$?
$x < -1$	$x = -2$	$(-2 + 1)(-2 - 3) = 5$	$5 < 0$ is false.
$-1 < x < 3$	$x = 0$	$(0 + 1)(0 - 3) = -3$	$-3 < 0$ is true.
$x > 3$	$x = 4$	$(4 + 1)(4 - 3) = 5$	$5 < 0$ is false.

Since the factored form of the inequality is true for a value ($x = 0$) in the interval $-1 < x < 3$, it must also be true for every number in this interval. Hence, the solution to

$$x^2 - 2x < 3$$

is

$$-1 < x < 3.$$

General form of the solution intervals: If r_1 and r_2 $(r_1 < r_2)$ represent the roots of $ax^2 + bx + c = 0$ $(a \neq 0)$, then the solution intervals to the related set of quadratic inequalities are given in the table. Knowing these general forms can greatly simplify matters. For example, the solution interval to $x^2 + x - 6 \leq 0$ has the general form $r_1 \leq x \leq r_2$ where r_1 and r_2 are the roots of the related quadratic equation $x^2 + x - 6 = 0$. Since the roots of this equation are $r_1 = -3$ and $r_2 = 2$, the solution to $x^2 + x - 6 \leq 0$ is $-3 \leq x \leq 2$. If the original quadratic inequality was $x^2 + x - 6 > 0$, then the solution would be $x < -3$ or $x > 2$.

General Solutions to Quadratic Inequalities ($a > 0$)

Inequality	Solution Interval(s)	Graph
$ax^2 + bx + c < 0$	$r_1 < x < r_2$	
$ax^2 + bx + c \leq 0$	$r_1 \leq x \leq r_2$	
$ax^2 + bx + c > 0$	$x < r_1$ or $x > r_2$	
$ax^2 + bx + c \geq 0$	$x \leq r_1$ or $x \geq r_2$	

Key 15 Radical equations

OVERVIEW *An equation in which the variable appears under a radical sign is solved by isolating the radical and then raising each side of the equation to the power that eliminates the radical.*

Extraneous roots: Performing the same algebraic operation on both sides of an equation does not always produce an equivalent equation. Roots of the new equation that are not roots of the original equation are called *extraneous roots*. Extraneous roots may arise when both sides of an equation are multiplied or divided by the same variable expression, or when both sides of an equation are raised to the same power. For example, squaring both sides of the equation $x = 3$ produces the equation $x^2 = 9$ whose roots are -3 and 3. Since -3 is not a root of the original equation, it is an extraneous root.

Radical equation: In this type of equation, the variable appears underneath a radical sign. A radical equation such as

$$\sqrt{2 - x} - x = 0$$

is solved by isolating the radical and then raising both sides of the equation to the power that eliminates the radical. Squaring both sides of the equation

$$\sqrt{2 - x} = x$$

gives

$$2 - x = x^2$$

or, equivalently,

$$x^2 + x - 2 = 0$$

whose roots are -2 and 1. Checking both roots in the original equation shows that $x = 1$ is a root of the radical equation, but $x = -2$ is an extraneous root.

To solve $x^{p/k} = c$: Raise both sides of the equation to the power that is the reciprocal of the fractional exponent.

Example 1. If $x^{2/3} = -4$, then $x = (-4)^{3/2} = \sqrt{-64}$.

Since $\sqrt{-64}$ is not real, the original equation does not have a real solution.

Example 2. If
$$x^{3/5} + 9 = 1,$$
then
$$x^{3/5} = -8,$$
so

$$x = (-8)^{5/3} = (\sqrt[3]{-8})^5 = (-2)^5 = -32.$$

KEY EXAMPLE

Solve for x and check: $\sqrt{x+4} + \sqrt{1-x} = 3$

Solution: Isolate one radical term and then eliminate this radical by raising both sides of the equation to the second power:

$$\sqrt{x+4} = 3 - \sqrt{1-x}$$

$$(\sqrt{x+4})^2 = (3 - \sqrt{1-x})^2$$

$$x+4 = 9 - 6\sqrt{1-x} + (1-x)$$

Isolate the remaining radical term and then raise both sides of the equation to the second power:

$$(2x-6)^2 = (-6\sqrt{1-x})^2$$

$$4x^2 - 24x + 36 = 36(1-x)$$

$$4x^2 + 12x = 0$$

$$4x(x+3) = 0$$

Since $4x = 0$ or $x + 3 = 0$, $x = 0$ or $x = -3$. The solution is $x = 0$ or $x = -3$. You should verify that both solutions satisfy the original radical equation.

KEY EXAMPLE

Solve for x: $\sqrt{(x-1)^3} - 6 = 2$

Solution: If

$$\sqrt{(x-1)^3} - 6 = 2,$$

then

$$\sqrt{(x-1)^3} = 8.$$

Square each side of the equation:

$$(x-1)^3 = 64.$$

Raise each side of the equation to the $\frac{1}{3}$ power:

$$x - 1 = 64^{\frac{1}{3}} = 4$$

Since $x - 1 = 4$, $x = 5$. The check is left for you.

KEY EXAMPLE

Solve for y and check:

$$2\left(\frac{1}{y+1}\right)^{\frac{2}{3}} - 5\left(\frac{1}{y+1}\right)^{\frac{1}{3}} + 2 = 0$$

Solution: Make a change in the variable by letting

$$u = \left(\frac{1}{y+1}\right)^{\frac{1}{3}}$$

so that

$$u^2 = \left(\frac{1}{y+1}\right)^{\frac{2}{3}}.$$

Thus, the original equation becomes

$$2u^2 - 5u + 2 = 0,$$

whose roots are $u = \frac{1}{2}$ and $u = 2$. If

$$u = \left(\frac{1}{y+1}\right)^{\frac{1}{3}} = \frac{1}{2},$$

then raising both sides of this equation to the third power gives

$$\frac{1}{y+1} = \frac{1}{8}, \text{ so } y = 7$$

If $u = \left(\dfrac{1}{y+1}\right)^{\frac{1}{3}} = 2$, then:

$$\frac{1}{y+1} = 2^3 = 8, \text{ so } y = -\frac{7}{8}.$$

The roots of the original equation are $y = 7$ and $y = -\dfrac{7}{8}$. The check is left for you.

Theme 3 RATIONAL EXPRESSIONS

*I*f the numerator and the denominator of a fraction are polynomials, then the fraction is called a **rational expression**. We always assume that whenever a variable appears in the denominator of a fraction, it cannot have a value that will make the denominator evaluate to 0. The rules for working with fractions in algebra parallel the rules for operating with fractions in arithmetic.

INDIVIDUAL KEYS IN THIS THEME

---

16 Simplifying rational expressions

---

17 Operations with rational expressions

---

18 Simplifying complex fractions

---

19 Fractional equations and inequalities

Key 16 Simplifying rational expressions

OVERVIEW *Rational expressions are simplified by dividing the numerator and the denominator by any factors common to both.*

Negative of a fraction: The negative of $\frac{a}{b}$ may be written in the following equivalent ways:

$$-\frac{a}{b}, \frac{-a}{b}, \text{ and } \frac{a}{-b}.$$

Cancellation law: The cancellation law may be expressed in the following way:

$$\frac{ac}{bc} = \frac{a \cdot \overset{1}{\cancel{c}}}{b \cdot \cancel{c}} = \frac{a}{b} \cdot 1 = \frac{a}{b} \, (b, c \neq 0).$$

To simplify a rational expression: Factor the numerator and the denominator completely. Use the cancellation law to eliminate any factor that is common to both the numerator and the denominator since the quotient of like factors is 1. Then multiply the remaining factors. Sometimes it is necessary to factor out –1 from a factor in order to cancel out like factors, as in

$$\frac{2a - 2b}{bx - ax} = \frac{2(a - b)}{x(b - a)} = \frac{-2(\overset{1}{\cancel{b - a}})}{x(\cancel{b - a})} = -\frac{2}{x}.$$

KEY EXAMPLE

Simplify:
$$\frac{n - m}{m^2 - n^2}$$

Solution: Factor the denominator as the difference of two squares, $(m - n)(m + n)$. By factoring out –1 from the numerator, the quotient of a pair of like factors can be obtained.

$$\frac{n - m}{m^2 - n^2} = \frac{n - m}{(m - n)(m + n)} = \frac{-1(\overset{1}{\cancel{m - n}})}{(\cancel{m - n})(m + n)} = -\frac{1}{m + n}.$$

KEY EXAMPLE

Write the fraction $\dfrac{x^3 - 9x}{x^2 + 4x + 3}$ in lowest terms.

Solution: Factor the numerator completely and factor the denominator as the product of two binomials.

$$\frac{x^3 - 9x}{x^2 + 4x + 3} = \frac{x(x^2 - 9)}{(x + 3)(x + 1)}$$

$$= \frac{x\cancel{(x + 3)}^{1}(x - 3)}{\cancel{(x + 3)}(x + 1)}$$

$$= \frac{x^2 - 3x}{x + 1}$$

Key 17 Operations with rational

expressions

OVERVIEW *The rules for multiplying, dividing, and combining rational expressions are the same as those for fractions in arithmetic.*

To multiply rational expressions: If possible, factor the numerators and the denominators so that, if the same factor appears in both a numerator and a denominator, it can be divided out before multiplying. Write the product of the factors remaining in the numerators over the product of the factors remaining in the denominators. The result is the product of the original fractions in lowest terms.

KEY EXAMPLE

Write the product in lowest terms:

$$\frac{x^2 - 16}{10x} \cdot \frac{15x^3}{3x + 12}$$

Solution: Factor the numerators and the denominators.

$$\frac{x^2 - 16}{10x} \cdot \frac{15x^3}{3x + 12} = \frac{(x - 4)(x + 4)}{(5x)(2)} \cdot \frac{(5x)(3x^2)}{3(x + 4)}$$

$$= \frac{(x - 4)}{2} \cdot \frac{x^2}{1}$$

$$= \frac{x^3 - 4x^2}{2}$$

To divide rational expressions: Invert the second fraction and then multiply the two fractions.

KEY EXAMPLE

Write each quotient in lowest terms:

(a) $\dfrac{6x - 3x^2}{9x^3 - x} \div \dfrac{x^2 - 4}{3x^2 + 5x - 2}$ (b) $\dfrac{x^2 + 2xy + y^2}{xy - x} \div \dfrac{x^2 - y^2}{y^2 - y}$

Solution: Change to an equivalent multiplication example by inverting the second fraction.

(a) $\dfrac{6x - 3x^2}{9x^3 - x} \div \dfrac{x^2 - 4}{3x^2 + 5x - 2} = \dfrac{6x - 3x^2}{9x^3 - x} \cdot \dfrac{3x^2 + 5x - 2}{x^2 - 4}$

$$= \dfrac{3\overset{1}{\cancel{x}}\,(2 \overset{-1}{\cancel{-x}})}{\cancel{x}(\cancel{3x-1})(3x+1)} \cdot \dfrac{\overset{1}{(\cancel{3x-1})}\,\overset{1}{(\cancel{x+2})}}{(\cancel{x-2})(\cancel{x+2})}$$

$$= \dfrac{-3}{3x+1}$$

(b) $\dfrac{x^2 + 2xy + y^2}{xy - x} \div \dfrac{x^2 - y^2}{y^2 - y} = \dfrac{x^2 + 2xy + y^2}{xy - x} \cdot \dfrac{y^2 - y}{x^2 - y^2}$

$$= \dfrac{(x + y)\overset{1}{(\cancel{x+y})}}{x(\cancel{y-1})} \cdot \dfrac{y\overset{1}{(\cancel{y-1})}}{(\cancel{x+y})(x - y)}$$

$$= \dfrac{(x + y)y}{x(x - y)}$$

$$= \dfrac{xy + y^2}{x^2 - xy}$$

To combine rational expressions: If the fractions have like denominators, write the sum (or difference) of their numerators over their common denominator and then simplify, if possible. If the fractions have unlike denominators, change the fractions to equivalent fractions having the lowest common multiple of their denominators as their common denominator. This denominator is called the **LCD** (lowest common denominator).

KEY EXAMPLE

Express the sum in lowest terms:

$$\frac{4x}{x^2 - 4} + \frac{x+6}{4 - x^2}$$

Solution: Obtain like denominators by multiplying the numerator and the denominator of the second fraction by -1:

$$\frac{4x}{x^2 - 4} + \frac{-(x+6)}{-(4-x^2)} = \frac{4x}{x^2-4} + \frac{-(x+6)}{x^2-4}$$

$$= \frac{4x - (x+6)}{x^2 - 4}$$

$$= \frac{3x - 6}{x^2 - 4}$$

$$= \frac{3(\overset{1}{\cancel{x-2}})}{(\cancel{x-2})(x+2)}$$

$$= \frac{3}{x+2}$$

KEY EXAMPLE

Combine and express the result in lowest terms:

$$\frac{x}{x^2 + x - 6} - \frac{2}{x+3} + \frac{x-3}{x^2 - 4x + 4}$$

Solution: Factor the denominators of the first and last fractions:

$$x^2 + x - 6 = (x+3)(x-2)$$

and

$$x^2 - 4x + 4 = (x-2)^2$$

The LCD includes all the different prime factors of the three denominators with each prime factor raised to the highest power to which it occurs in any denominator. Since the factor $(x - 2)$ is raised to the second power in the last denominator, the LCD $= (x + 3)(x - 2)^2$. Multiply each fraction by "1" in the form of the factor(s) of the LCD its denominator lacks, divided by the same factor(s).

Thus

$$\frac{x}{x^2 + x - 6} - \frac{2}{x + 3} + \frac{x - 3}{x^2 - 4x + 4}$$

$$= \frac{x}{(x + 3) \cdot (x - 2)} \cdot \frac{(x - 2)}{(x - 2)} - \frac{2}{x + 3} \cdot \frac{(x - 2)^2}{(x - 2)^2} + \frac{x - 3}{(x - 2)^2} \cdot \frac{(x + 3)}{(x + 3}$$

$$= \frac{x(x - 2)}{(x + 3)(x - 2)^2} - \frac{2(x - 2)^2}{(x + 3)(x - 2)^2} + \frac{(x - 3)(x + 3)}{(x - 2)^2(x + 3)}$$

$$= \frac{x^2 - 2x - 2(x^2 - 4x + 4) + x^2 - 9}{(x + 3)(x - 2)^2}$$

$$= \frac{6x - 17}{(x + 3)(x - 2)^2}$$

Key 18   Simplifying complex fractions

OVERVIEW   *A **complex fraction** is a fraction that contains other fractions in its numerator or denominator, or in both. Simplifying a complex fraction means eliminating any fractions from its numerator and denominator.*

To simplify a complex fraction by combining terms: Rewrite the complex fraction so that a single fraction appears in its numerator and in its denominator. Then divide the numerator of the complex fraction by its denominator.

Example:

$$\frac{n - \dfrac{1}{n}}{\dfrac{1}{n} + 1} = \frac{\dfrac{n^2 - 1}{n}}{\dfrac{1 + n}{n}} = \frac{n^2 - 1}{n} \div \frac{1 + n}{n}$$

$$= \frac{n^2 - 1}{n} \times \frac{n}{1 + n}$$

$$= \frac{(n - 1)(n + 1)}{n} \times \frac{n}{1 + n} = n - 1$$

To simplify a complex fraction by clearing fractions: Multiply each term in both the numerator and the denominator of the complex fraction by the LCD of all the fractions that the numerator and the denominator contains.

Example:

$$\frac{ab^{-1} + 1}{ab^{-1} - a^{-1}b} = \frac{\dfrac{a}{b} + 1}{\dfrac{a}{b} - \dfrac{b}{a}}$$

LCD = ab:

$$= \frac{(ab)}{(ab)} \left(\frac{\dfrac{a}{b} + 1}{\dfrac{a}{b} - \dfrac{b}{a}} \right) = \frac{(ab)\left(\dfrac{a}{b} + 1 \right)}{(ab)\left(\dfrac{a}{b} - \dfrac{b}{a} \right)}$$

$$= \frac{a^2 + ab}{a^2 - b^2} \qquad = \frac{a(a + b)}{(a + b)(a - b)} = \frac{a}{a - b}$$

Key 19 Fractional equations and inequalities

OVERVIEW *Equations and inequalities in which the variable appears in the denominator of a fraction are solved by clearing the fractions using the multiplication properties of equations or inequalities.*

To solve a fractional equation: Multiply each term of the equation, right and left sides, by the LCD. The solution of the transformed equation will include all the roots of the fractional equation, but may also include extraneous roots.

KEY EXAMPLE

Solve and check:

$$\frac{x}{x+2} - \frac{1}{x-2} = \frac{8}{x^2 - 4}$$

Solution: The last denominator in factored form is $(x - 2)(x + 2)$. Eliminate each of the denominators by multiplying each member of the equation by the LCD, $(x - 2)(x + 2)$. Thus,

$$\frac{x}{\cancel{x+2}}[(x - 2)(\cancel{x+2})] - \frac{1}{\cancel{x-2}}[(\cancel{x-2})(x + 2)]$$

$$= \frac{8}{\cancel{(x-2)(x+2)}}[\cancel{(x-2)(x+2)}]$$

$$x(x - 2) - (x + 2) = 8$$

$$x^2 - 2x - x - 2 - 8 = 0$$

$$x^2 - 3x - 10 = 0$$

$$(x - 5)(x + 2) = 0$$

If $x - 5 = 0$, $x = 5$. If $x + 2 = 0$, then $x = -2$. Since $x = -2$ makes the denominator of the first fraction in the original equation 0, this root is an extraneous root and must be rejected. Since $x = 5$ satisfies the fractional equation, it is a root.
The solution is $x = 5$.

To solve fractional inequality: To solve a fractional inequality such as

$$\frac{x-4}{x+2} \ge 0,$$

determine the intervals on which the fraction takes on nonnegative values. Begin by locating on a number line the *critical* x-values for which the fraction is undefined or 0. The fraction is undefined when $x = -2$ and evaluates to 0 when $x = 4$. Indicate on the number line that -2 does not belong to the solution set by enclosing -2 it in an open circle. Use a closed circle around 4 to show that $x = 4$ is a member of the solution set. See Figure 3.1.

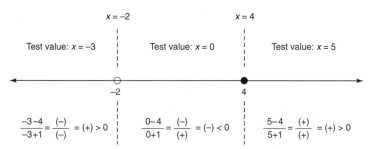

Figure 3.1 Using Test Values to Find Solution Intervals

Use a test value that belongs to each of the three intervals to determine whether the inequality is true on that interval, as illustrated in Figure 3.1. The solution intervals are $(-\infty, -2)$ and $(4, \infty)$, which may be written also as $x < -2$ and $x \ge 4$.

If the original inequality is $\le$ or $\ge$, include in the final solution each x-value that makes the numerator 0; otherwise, do not include those numbers in the solution. Always exclude from the solution each x-value for which the denominator is 0.

KEY EXAMPLE

Solve:

$$\frac{5}{x-3} \le \frac{2}{x}$$

Solution: Rewrite the inequality so that 0 is on the right side and a single fraction is on the left side:

$$\frac{5}{x-3} - \frac{2}{x} \le 0$$

$$\frac{5x - 2(x-3)}{x(x-3)} \le 0$$

$$\frac{3(x+2)}{x(x-3)} \le 0$$

- Locate the critical numbers –2, 0, and 3 on a number line. Place a closed circle around –2 since the fraction can be equal to 0. Use open circles around 0 and 3 since for these values of x the fraction is undefined.

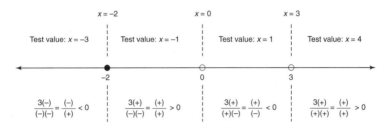

- Using appropriate test values, determine the sign of the fraction on each of the four intervals determined by the three critical numbers, as shown in the accompanying figure.
- Summarize the solution intervals: $(-\infty, 2)$ and $(0, 3)$, which may be written also as $x \leq -2$ and $0 < x < 3$.

Finding solutions intervals on a number line: When the numerator and denominator consist of distinct linear factors in x or odd powers of x, the critical numbers divide the number line so that every other interval belongs to the solution. This fact can help you quickly determine solution intervals on a number line. In the Key Example above, once you determine that the solution includes the leftmost interval, $x \leq -2$, you need not do any additional work because the solution intervals alternate thereafter, as shown in Figure 3.2.

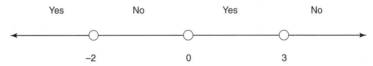

Figure 3.2 Alternating Solution Intervals

If the leftmost interval did not belong to the solution, the solution would alternate beginning with the next consecutive interval.

Theme 4 LINEAR EQUATIONS
AND INEQUALITIES
IN TWO VARIABLES

*T*he French mathematician René Descartes (1596–1650) developed a method for representing equations in two variables graphically. This method is based on creating a **coordinate plane** by drawing a horizontal number line called the **x-axis** and a vertical number line called the **y-axis**. The point at which the two coordinate axes intersect is called the **origin**. Each point in the plane corresponds to an ordered pair of real numbers, (x, y). By locating points in the plane that satisfy a given equation, we can draw its graph. Every first-degree equation has a line as its graph; and, conversely, every line can be described by a first-degree equation. The line $ax + by = c$ divides the coordinate plane into two half-planes, one of which contains all points that satisfy the inequality $ax + by > c$; the other half-plane contains all points that satisfy the inequality $ax + by < c$.

Key 20 Midpoint and distance formulas

OVERVIEW *If the endpoints of a line segment are known, then convenient formulas can be used to determine the midpoint and length of the line segment.*

Cartesian coordinate system: In the Cartesian coordinate system, ordered pairs of real numbers of the form (x, y) represent the locations of points from two perpendicular axes that intersect at the **origin**. The first member of the ordered pair (x, y), called the **x-coordinate** or **abscissa**, tells the number of units that the point is located horizontally to the right $(x > 0)$ or to the left $(x < 0)$ of the origin. The second member of the ordered pair (x, y), called the **y-coordinate** or **ordinate**, gives the number of units that the point is located vertically above $(y > 0)$ or below $(y < 0)$ the origin. In Figure 4.1 point $A(2, 3)$ is graphed by starting at the origin and moving along the x-axis 2 units to the right, stopping, and then moving 3 units up. Similarly, point $B(-4, 5)$ is graphed by starting at the origin and moving along the x-axis 4 units to the *left* (since $x < 0$), and then moving up 5 units.

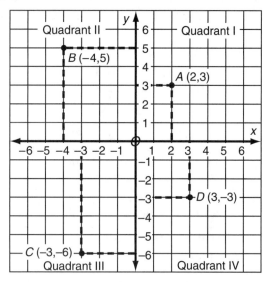

Figure 4.1 Graphing Points in the Four Quadrants

The four quadrants: The coordinate axes divide the coordinate plane into four distinct regions called **quadrants.** In Figure 4.1, the four quadrants are numbered I, II, III, and IV in counterclockwise order, beginning in the upper right corner. The signs of the x- and y-coordinates of a point determine the quadrant in which the point lies, as shown in the table that follows.

Example	Signs	Quadrant
$A(2, 3)$	$(+, +)$	I
$B(-4, 5)$	$(-, +)$	II
$C(-3, -6)$	$(-, -)$	III
$D(3, -3)$	$(+, -)$	IV

Midpoint formula: If $M(\bar{x}, \bar{y})$ is the midpoint of the line segment whose endpoints are $A(x_1, y_1)$ and $B(x_2, y_2)$, as shown in Figure 4.2, then

$$M(\bar{x}, \bar{y}) = \left(\frac{x_1 + x_2}{2}, \frac{y_1 + y_2}{2} \right).$$

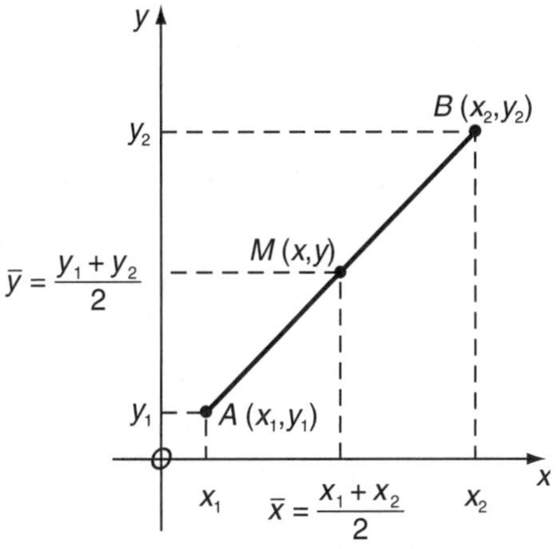

Figure 4.2 Midpoint of $\overline{AB}$

Distance formula: The distance d between points $A(x_1, y_1)$ and $B(x_2, y_2)$, as shown in Figure 4.3, is given by the equation

$$d = \sqrt{(x_2 - x_1)^2 + (y_2 - y_1)^2}.$$

If points A and B lie on the same *vertical* line, then $x_2 = x_1$ and $d = |y_2 - y_1|$. If points A and B lie on the same *horizontal* line, then $y_2 = y_1$ and $d = |x_2 - x_1|$.

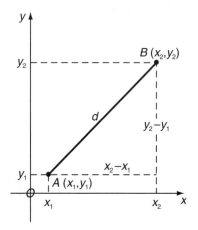

Figure 4.3 Length of $\overline{AB}$

KEY EXAMPLE

The coordinates of the endpoints of a line segment are $A(-1, 9)$ and $B(5, 1)$. Find the midpoint M and length d of $\overline{AB}$.

Solution: Let $(x_1, y_1) = (-1, 9)$ and $(x_2, y_2) = (5, 1)$.

$$M(x, y) = \left(\frac{x_1 + x_2}{2}, \frac{y_1 + y_2}{2} \right) \qquad d = \sqrt{(x_2 - x_1)^2 + (y_2 - y_1)^2}$$

$$= \left(\frac{(-1) + 5}{2}, \frac{9 + 1}{2} \right) \qquad = \sqrt{[5 - (-1)]^2 + (1 - 9)^2}$$

$$= (2, \ 5) \qquad = \sqrt{36 + 64} = \sqrt{100} = 10$$

You should verify that, by letting $(x_1, y_1) = (5, 1)$ and $(x_2, y_2) = (-1, 9)$, the same results are obtained.

KEY EXAMPLE

The center of a circle is located at (7, –1) and the coordinates of one of the endpoints of a diameter of this circle are (5, 4). What are the coordinates of the other endpoint of this diameter?

Solution: The center of a circle is the midpoint of each of its diameters. Use the midpoint formula where (x, y) represents the coordinates of the unknown endpoint. Since

$$(7,-1) = \left(\frac{5+x}{2}, \frac{4+y}{2} \right),$$

$$7 = \frac{5+x}{2} \quad \text{and} \quad -1 = \frac{4+y}{2}.$$

Multiplying each side of each equation by 2 gives

$$14 = 5 + x \quad \text{and} \quad -2 = 4 + y.$$

Thus, $x = 9$ and $y = -6$. The coordinates of the other endpoint of the diameter are (9, –6).

Key 21 Slope formula

OVERVIEW *Moving from one point to another point on an oblique (slanted) line produces a vertical change in y-coordinates and a horizontal change in x-coordinates. The ratio of these changes provides a measure of the steepness or **slope** of the line.*

Slope formula: The slope m of a nonvertical line that contains the points (x_1, y_1) and (x_2, y_2) as shown in Figure 4.4, is defined by the following equation:

$$m = \frac{\text{vertical change in } y}{\text{horizontal change in } x} = \frac{y_2 - y_1}{x_2 - x_1}$$

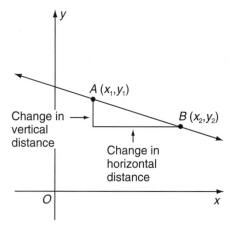

Figure 4.4 Slope of a Line

As the slope of a line increases in absolute value, the line gets "steeper."

Example: To find the slope of the line that contains points $(1, -2)$ and $(4, 7)$, let $(x_1, y_1) = (1, -2)$ and $(x_2, y_2) = (4, 7)$. Thus,

$$m = \frac{y_2 - y_1}{x_2 - x_1} = \frac{7 - (-2)}{4 - 1} = \frac{9}{3} = 3$$

You should verify that the same slope is obtained by considering (4, 7) as the "first" point and letting $(x_1, y_1) = (4, 7)$ and $(x_2, y_2) = (1, -2)$.

Positive and negative slopes: If a line rises as x increases, then the slope m of the line is positive. A line has a negative slope if the line falls as x increases. These situations are shown in Figure 4.5.

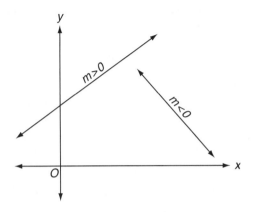

Figure 4.5 Positive and Negative Slopes

Zero slope: The slope of a horizontal line is zero.

Undefined slope: The slope of a vertical line is not defined.

Slopes of parallel lines: Parallel lines have the same slope. Conversely, different nonvertical lines that have the same slope are parallel.

Slopes of perpendicular lines: Perpendicular lines have slopes whose product is −1. Conversely, lines whose slopes have a product of −1 are perpendicular.

Key 22 Equations of lines

OVERVIEW *The numerical relationship between the* x-
and y-*coordinates of all points on a line is the same and can
be expressed as a first-degree equation in* x *and* y.

Equation of a line in standard form: If a, b, and c are constants and
a and b are not both zero, then $ax + by = c$ represents the standard
form of the equation of a line.

Intercepts: An oblique line will cross both the x- and y-axis. The x-
coordinate of the point at which a line crosses the x-axis is called the
***x*-intercept**. At this point, $y = 0$. The y-coordinate of the point at
which a line crosses the y-axis is called the **y-intercept.** At this
point, $x = 0$.

Example: If an equation of a line is $3x - 4y = 12$, then replacing y
with 0 and solving the resulting equation for x yields $x = 4$ as the
x-intercept. Similarly, replacing x with 0 in $3x - 4y = 12$ and solving
for y identifies $y = -3$ as the y-intercept.

Equation of a line in point-slope form: If m is the slope of a nonvertical line and (x_1, y_1) is a point on the line, then an equation of this
line is $y - y_1 = m(x - x_1)$. For example, if a line whose slope is 2
contains the point $(-3, 1)$, then an equation of this line is found by
letting $m = 2$ and $(x_1, y_1) = (-3, 1)$:

$$y - y_1 = m(x - x_1)$$
$$y - 1 = 2(x - (-3))$$
$$y = 2x + 7$$

Equation of a line in slope-intercept form: If m represents the slope
of a nonvertical line and b is its y-intercept, then an equation of this
line is $y = mx + b$. If a line has a slope of -1 and a y-intercept of 4,
then an equation of this line is $y = -x + 4$.

Example: To find the slope and y-intercept of the line

$$10x + 2y = 5,$$

put the equation in $y = mx + b$ form:

$$y = -5x + \frac{5}{2}.$$

Since $m = -5$ and $b = \dfrac{5}{2}$, the slope of the line is -5 and its y-intercept is $\dfrac{5}{2}$.

Comparing equations of lines: Comparing the slopes of lines whose equations are in slope-intercept form may provide useful information about the lines.

Example 1: The lines $y = 2x - 1$ and $y = 2x + 3$ have the same slope ($m = 2$) and, as a result, are parallel.

Example 2: The lines $y = -3x$ and $y = \dfrac{1}{3}x + 7$ have slopes whose product is

$$-3\left(\frac{1}{3}\right) = -1$$

and, as a result, are perpendicular.

KEY EXAMPLE

Write an equation of a line that is parallel to the line $y + 2x = 5$ and contains point $(-1, 2)$.

Solution: In slope-intercept form, the given equation is $y = -2x + 5$ so the slope of its line is -2. Since parallel lines have the same slope, the slope of the desired line must also be -2. Use the point-slope form where $m = -2$ and $(x_1, y_1) = (-1, 2)$.

$$\begin{aligned} y - y_1 &= m(x - x_1) \\ y - 2 &= -2(x - (-1)) \\ y &= -2x \end{aligned}$$

KEY EXAMPLE

Write an equation of the line that contains points $(6, 0)$ and $(2, -6)$.

Solution: First find the slope of the line. Then use the point-slope form, letting (x_1, y_1) equal either of the given points.

$$m = \frac{-6 - 0}{2 - 6} = \frac{-6}{-4} = \frac{3}{2}$$
$$y - y_1 = m(x - x_1)$$

Let $(x_1, y_1) = (6, 0)$:
$$y - 0 = \frac{3}{2}(x - 6)$$
$$y = \frac{3}{2}x - 9$$

Equations of horizontal and vertical lines: A vertical line will intersect the x-axis and a horizontal line will intersect the y-axis.

- $x = a$ is an equation of a vertical line whose x-intercept is a.
- $y = b$ is an equation of a horizontal line whose y-intercept is b.

Example: An equation of a line that contains $(-3, 5)$ and is parallel to the x-axis is $y = 5$. If the line contains $(-3, 5)$ and is parallel to the y-axis, then an equation of the line is $x = -3$.

Key 23 Graphing linear equations

in two variables

OVERVIEW *Graphing a line means using several points that the line contains to place the line correctly on a coordinate plane.*

To graph a line from its equation:

1. Draw and label the coordinate axes on graph paper.
2. Choose an appropriate scale along each axis.
3. Determine the coordinates of at least two points that satisfy the equation of the line.
4. Graph these points, and then connect them with a straight line;
5. Label the line with its equation.

Determining points that satisfy a linear equation: Any of the following methods can be used to obtain points that satisfy a linear equation whose graph is an oblique (slanted) line:

- Intercept Method: Determine from the linear equation the point at which the line crosses the *x*-axis by letting $y = 0$ and solving for *x*, and the point at which the line crosses the *y*-axis by letting $x = 0$ and solving for *y*. This method can be used only for oblique lines that do not pass through the origin.

  Example: The graph of the equation $2x + 3y = 6$ crosses the *x*-axis at $(3, 0)$, and the *y*-axis at $(0, 2)$.

- Slope-Intercept Method: Put the equation in $y = mx + b$ form. Graph $(0, b)$, the point at which the line crosses the *y*-axis. Use the slope to locate at least one additional point.

  Example: The graph of the equation $y = 3x + 1$ crosses the *y*-axis at $(0, 1)$. Since the slope of the line is $3 \left(= \dfrac{1}{3} \right)$, you can obtain another point on this line by starting at $(0, 1)$ and moving vertically up 3 units and then moving horizontally to the right 1 unit. The coordinates of this point are $(0 + 1, 1 + 3) = (1, 4)$.

- Three-Point Method: Solve the given equation for *y*, if necessary. Determine the coordinates of three points on the line by replacing *x* with three convenient values and then determining the corresponding values of *y*. Although two points determine a line, the third point serves as a "check" point. This method is illustrated in the following Key Example.

KEY EXAMPLE

Sketch the graph of $y - 2x = 4$.

Solution:

1. Solve the given equation for *y*: $y = 2x + 4$.
2. Find the corresponding values of *y* for three convenient values of *x*, say $x = -2$, 0, and 2.

x	$2x + 4 = y$	(x, y)
–2	$2(-2) + 4 = 0$	$(-2, 0)$
0	$2(0) + 4 = 4$	$(0, 4)$
2	$2(2) + 4 = 8$	$(2,8)$

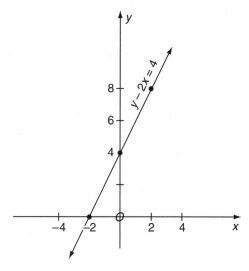

3. Graph (–2, 0), (0, 4), and (2, 8), and draw a straight line through them. If one of the points does not lie on the line, then at that point, the coordinates are not correct or the line has been graphed incorrectly.
4. Label the line with its equation.

Key 24 Solving a linear system
of equations

OVERVIEW *Solving a system of equations means finding the set of all ordered pairs of numbers that make each equation true at the same time. A system of linear equations can be solved either graphically or algebraically.*

Solving a linear system graphically: To solve the system

$$2x + y = 5$$
$$-x + y = 2$$

graphically, graph both equations on the same set of axes. The intercepts of $2x + y = 5$ are $\left(\frac{5}{2}, 0\right)$ and $(0,5)$. The intercepts of $y - x = 2$ are $(-2, 0)$ and $(0, 2)$. Plot each pair of intercepts on the same set of axes, and then draw the lines that contain them, as shown in Figure 4.6. The solution is $(1, 3)$, which can be checked algebraically by verifying that, when $x = 1$ and $y = 3$, both of the original equations are true.

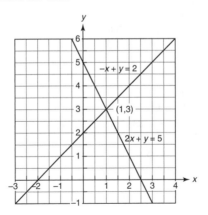

Figure 4.6 Graphs of $2x + y = 5$ and $-x + y = 2$.

Solving a linear system graphically with a calculator: To use a graphing calculator such as Texas Instruments model TI-84 Plus to solve the system of linear equations shown in Figure 4.6, first solve each equation for y. Set $Y_1 = -2x + 5$ and $Y_2 = x + 2$. After displaying the graphs in an appropriate window such as $[-4.7, 4.7] \times [-3.1, 6.2]$, move the cursor close to the point of intersection. Use either the TRACE feature or the **intersect** feature from the CALC menu to find the coordinates of the point of intersection. See Figures 4.7 and 4.8.

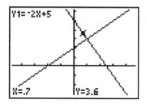

Figure 4.7 Using the **Intersect** Feature

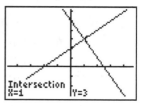

Figure 4.8 The Solution Is (1, 3).

The graphing calculator procedures described in this book will always refer to Texas Instruments model TI-84 Plus. If you have a different calculator, you may have to modify these procedures.

Solving a linear system algebraically: To solve a system of linear equations algebraically, reduce the system to a single equation with only one variable. To solve the system

$$2x + y = 5$$
$$-x + y = 2$$

algebraically, use either of the following methods:

• Substitution Method. To eliminate y in the first equation, first solve the second equation for y: $y = x + 2$. Then substitute $x + 2$ for y in the first equation. Thus,

$$2x + (x + 2) = 5, \text{ so } 3x + 2 = 5 \text{ and } x = \frac{3}{3} = 1.$$

When $x = 1$, $y = x + 2 = 1 + 2 = 3$. The solution is (1, 3).

• Addition Method. If necessary, write each equation in the form $ax + by = c$. Multiply one or both of the equations by a suitable nonzero number in order to produce a system of equivalent equations in which one of the variables has opposite numerical coefficients. In the system given above, multiply each member of the second equation by 2. Then add corresponding sides of the two equations:

$$\begin{array}{r} 2x + y = 5 \\ + \quad -2x + 2y = 4 \\ \hline 3y = 9, \text{ so } y = \frac{9}{3} = 3 \end{array}$$

Substitute 3 for y in the second equation:

$$-x + 3 = 2, \text{ so } -x = -1 \text{ and } x = 1.$$

The solution is (1, 3).

Key 25   Graphing linear inequalities
in two variables

OVERVIEW   *To graph a two-variable linear inequality, first graph the corresponding equation. Then determine which side of the line contains the set of points that satisfy the original inequality.*

To graph a linear inequality: First graph the corresponding linear equation, which serves as a boundary line. Next, pick a convenient test point that lies on one side of the line. If the coordinates of the test point satisfy the inequality, then the solution contains all points on the same side of the line. Otherwise, the solution contains all points on the opposite side of the line. If the inequality relation is < or >, draw a broken (dashed) boundary line. If the inequality relation is ≤ or ≥, draw a solid boundary line which indicates that points on the line are also included in the solution.

KEY EXAMPLE

Graph $2y - x < 8$.

Solution: If $x = 0$, then $y = 4$, so the y-intercept is 4. If $y = 0$, then $x = -8$ so the x-intercept is -8. Draw a broken line since points on the line do not satisfy the inequality.

Pick $(0, 0)$ as a test point:

$$2y - x < 8$$
$$?$$
$$2(0) - 0 < 8$$
$$0 < 8 \quad \text{True.}$$

Since the test point satisfies the inequality, shade in the region on the side of the line where the point lies, as shown on the next page.

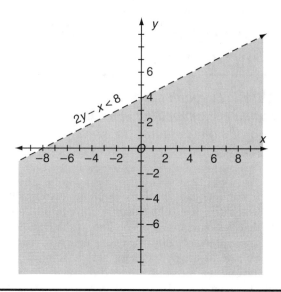

Solving a system of linear inequalities: To find the set of all ordered pairs that satisfy a system of two linear inequalities, graph each inequality on the same set of axes. Then determine the region in which the two solution sets overlap. For example, to find the solution of this system of:

$$y > 3x - 4$$
$$x + 2y \leq 6$$

follow these steps:

Step 1. Graph the boundary lines $y = 3x - 4$ and $x + 2y = 6$, as shown in Figure 4.9.

Step 2. For each inequality, decide whether the solution set lies above or below the boundary line. If you are not sure, pick a test point not on the boundary line.
- Consider $y > 3x - 4$ and test point (0, 0). Since $0 > 3 \cdot 0 - 4$ is true, the region on the side of the boundary line that includes (0, 0) represents the solution set of $y > 3x - 4$. Shade in this region, as shown in Figure 4.10.
- Consider $x + 2y \leq 6$ and test point (0, 0). Because $0 + 2(0) \leq 6$ is true, the region on the side of the boundary line that includes (0, 0) represents the solution set of $x + 2y \leq 6$. Shade in this region, as shown in Figure 4.10.

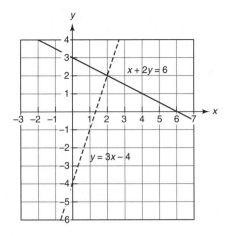

Figure 4.9 Graphing the Boundary Lines $y = 3x - 4$ and $x + 2y = 6$

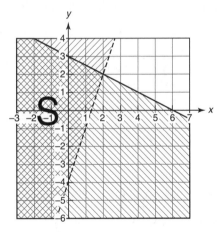

Figure 4.10 Identifying the Solution Set

Step 3. Identify the solution set. The solution set is the cross-hatched region in which the individual solution sets overlap. Label the solution set with an "**S**," as in Figure 4.10.

Theme 5 COMPLEX NUMBERS AND QUADRATIC EQUATIONS

A square root of a negative number is given meaning by defining a new type of number called an **imaginary number**. The set of **complex numbers** includes real numbers, imaginary numbers, and their sums. Quadratic equations that cannot be solved by factoring have irrational or imaginary roots that can be obtained using the **quadratic formula**. Every quadratic equation in two variables, only one of which is squared, has a graph called a **parabola**.

Key 26 Complex numbers

OVERVIEW *Although $\sqrt{-1}$ satisfies the equation* $x^2 = -1$, *the equation has no real roots. As a result, mathematicians expanded our number system by introducing the imaginary unit* i *and defining it so that* $i = \sqrt{-1}$ *and* $i^2 = -1$.

Pure imaginary numbers: Pure imaginary numbers have the form bi where b is a nonzero real number and $i = \sqrt{-1}$. Square roots of negative numbers other than -1 can be expressed as pure imaginary numbers by factoring out $\sqrt{-1}$ and replacing it with i.

Example 1. $\sqrt{-9} = \sqrt{9}\sqrt{-1} = 3i$

Example 2. $\sqrt{-50} = \sqrt{50}\sqrt{-1} = \sqrt{25}\sqrt{2}i = 5\sqrt{2}i$

To combine two imaginary numbers: Combine the numerical factors of i, as in $2i + 5i = 7i$.

To multiply two imaginary numbers: Use the rule

$$ci \cdot di = cd(i^2) = -cd$$

Example: $2i \cdot 5i = 10i^2 = -10$

To multiply two imaginary numbers that are in radical form, express each number in terms of i and then multiply.

Example: $\sqrt{-4} \cdot \sqrt{-25} = (2i)(5i) = 10i^2 = -10$

Notice that $\sqrt{-4} \cdot \sqrt{-25} \neq \sqrt{100}$. The rule $\sqrt{a} \cdot \sqrt{b} = \sqrt{ab}$ holds when at least one of a and b is nonnegative, but is not valid if a and b are both negative.

Simplifying powers of i: Any nonnegative integer power of i can be reduced to ± 1 or $\pm i$. We know that $i^0 = 1$, $i^1 = i$, $i^2 = -1$, and $i^3 = i^2 i = -i$. Higher integer powers of i follow the same cyclic pattern, which repeats every four integers. Thus, to evaluate i^n, where n is a positive integer greater than 3, divide n by 4. Using the remainder as the new power of i, simplify. For example, $i^{14} = i^3$ since dividing 14 by 4 gives a remainder of 3. Thus, $i^{14} = i^3 = -1$.

Complex number: A complex number is the sum of a real number and a pure imaginary number. The complex number $4 + 3i$ is in **standard form**, $a + bi$, where a and b are real numbers and $i = \sqrt{-1}$. Every real number is a complex number. For example, since 4 can be written as 4 + 0*i*, 4 is a complex number. Hence, the set of real numbers is a subset of the set of complex numbers.

Operations with complex numbers: Arithmetic operations with complex numbers are defined so that the field properties of real numbers hold also for the set of complex numbers. In performing arithmetic operations, complex numbers are treated like binomials, as in:

$$(2 + 3i) + (4 - 5i) = (2 + 4) + (3i - 5i) = 6 - 2i.$$

Two complex numbers can be multiplied using FOIL.
Example:

$$\underbrace{\quad}_{\textbf{F}} \qquad \underbrace{\quad}_{\textbf{O}} \qquad \underbrace{\quad}_{\textbf{I}} \qquad \underbrace{\quad}_{\textbf{L}}$$

$$(2 + 3i)(4 - 5i) = 2 \cdot 4 + (2)(-5i) + (3i)(4) + (3i)(-5i)$$
$$= 8 - 2i - 15i^2$$
$$= 8 - 2i - 15\,(-1)$$
$$= 23 - 2i$$

Complex conjugates: Complex conjugates are two complex numbers such as $a + bi$ and $a - bi$ that are the sum and difference of the same two terms. The product of a pair of complex conjugates is always a real number since

$$(a + bi)(a - bi) = (a^2) + (-abi) + (abi) + (-b^2)(i^2)$$
$$= a^2 + 0 \qquad + (-b^2)(-1)$$
$$= a^2 + b^2$$

Dividing complex numbers: To express the quotient of $\dfrac{c + di}{a + bi}$ as a complex number in standard form, multiply the fraction by 1 in the form of the complex conjugate of the denominator divided by itself. If $a = 0$, then multiply the numerator and the denominator by i.

KEY EXAMPLE

Express the quotients in standard form:

(a) $\dfrac{2 - 7i}{2i}$
(b) $\dfrac{5i}{1 + 4i}$

Solution:

(a) $\dfrac{2 - 7i}{2i} = \dfrac{2 - 7i}{2i} \cdot \dfrac{i}{i} = \dfrac{i(2 - 7i)}{2i^2}$

Multiply each term inside
the parentheses by i:

$$= \frac{2i - 7i^2}{2i^2}$$

Replace i^2 by -1:

$$= \frac{2i + 7}{-2}$$

Divide each term of the
numerator by -2:

$$= -\frac{7}{2} - i$$

(b) Multiply the numerator and the denominator by $1 - 4i$, which is the complex conjugate of $1 + 4i$:

$$\frac{5i}{1 + 4i} = \frac{5i}{1 + 4i} \cdot \frac{1 - 4i}{1 - 4i}$$

$$= \frac{5i(1 - 4i)}{1^2 + 4^2}$$

Multiply each term inside
the parentheses by $5i$:

$$= \frac{5i - 20i^2}{17}$$

$$= \frac{20}{17} + \frac{5i}{17}$$

Key 27 Solving quadratic equations
by completing the square

OVERVIEW *If the quadratic polynomial of a quadratic equation in standard form cannot be factored, then the quadratic equation can be solved by writing it in the form $(Ax + B)^2 = k$. The solutions for x are obtained by taking the square root of each side of the equation.*

Perfect square trinomials: The trinomial that results from squaring a binomial is a **perfect square trinomial**. Because $(x + k)^2 = x^2 + (2k)x + k^2$, $x^2 + (2k)x + k^2$ is a perfect square trinomial. The last term of the perfect square trinomial, k^2, can be obtained from the coefficient of the middle term, $2k$, by dividing the middle term by 2 and then squaring the result: $(2k \div 2)^2 = k^2$. To "complete the square" of $x^2 + 10x + \boxed{?}$ so that a perfect square trinomial results, add 25 because $(10 \div 2)^2 = 25$. Thus, $x^2 + 10x + \boxed{25} = (x + 5)^2$.

Solving quadratic equations by completing the square: A quadratic equation such as $4x^2 - 24x + 7 = 0$ has quadratic polynomials that cannot be factored with integers. Solve $4x^2 - 24x + 7 = 0$ by **completing the square**, as illustrated below.

$$4x^2 - 24x + 7 = 0$$

Step 1. Divide each term of the equation by 4, the coefficient of x^2.

$$x^2 - 6x + \frac{7}{4} = 0$$

Step 2. Subtract the constant term from both sides of the equation.

$$x^2 - 6 = -\frac{7}{4}$$

Step 3. Take one-half the coefficient of x and square it.

$$\left[\frac{1}{2}(-6)\right]^2 = 9$$

Step 4. Add the result obtained in Step 3 to both sides of the equation.

$$x^2 - 6x + 9 = -\frac{7}{4} + 9$$

Step 5. Factor the left side of the equation, and simplify the right side.

$$(x - 3)^2 = \frac{29}{4}$$

Step 6. Take the square root of each side of the equation, and solve for x.

$$x - 3 = \pm \frac{\sqrt{29}}{2}$$

$$x = 3 \pm \frac{\sqrt{29}}{2}$$

Step 7. Write the two roots. $x_1 = 3 + \frac{\sqrt{29}}{2}$ and $x_2 = 3 - \frac{\sqrt{29}}{2}$

Key 28 Solving quadratic equations
by formula

OVERVIEW   *Solving the quadratic equation* $ax^2 + bx + c$
$= 0$ $(a \neq 0)$ *using the method of completing the square gives*
a general formula that can be used to solve any quadratic
equation.

The quadratic formula: When this formula is used, each of the roots
of the quadratic equation $ax^2 + bx + c = 0$ $(a \neq 0)$ is expressed in terms
of the constants a, b, and c:

$$x = \frac{-b \pm \sqrt{b^2 - 4ac}}{2a} \quad (a \neq 0).$$

The quantity underneath the radical sign is called the **discriminant**
and may be denoted by D. Thus, $D = b^2 - 4ac$. Since the quadratic
formula can be used to find the roots of *any* quadratic equation in
standard form, it should be memorized.

KEY EXAMPLE

Solve for x: $2x^2 + 3x = 1$.

Solution: Put the equation into the standard form

$$2x^2 + 3x - 1 = 0.$$

Because $a = 2$, $b = 3$, and $c = -1$,

$$D = b^2 - 4ac = 3^2 - 4(2)(-) = 17.$$

Thus,

$$x = \frac{-b \pm \sqrt{D}}{2a} = \frac{-(3) \pm \sqrt{17}}{2(2)} = \frac{-3 \pm \sqrt{17}}{4}.$$

The two roots are

$$x_1 = \frac{-3 + \sqrt{17}}{4} \quad \text{and} \quad x_2 = \frac{-3 - \sqrt{17}}{4}.$$

KEY EXAMPLE

Solve for x: $x^2 + 5 = 2x$.

Solution: Put the equation into the standard form

$$x^2 - 2x + 5 = 0.$$

Because $a = 1$, $b = -2$, and $c = 5$,

$$D = b^2 - 4ac = -16.$$

Hence,

$$x = \frac{-b \pm \sqrt{D}}{2a} = \frac{-(2) \pm \sqrt{-16}}{2(1)} = \frac{2}{2} \pm \frac{4i}{2} = 1 \pm 2i.$$

The two roots are $x_1 = 1 + 2i$ and $x_2 = 1 - 2i$.

Describing the type of roots: Without solving a quadratic equation having the form $ax^2 + bx + c = 0$, the type of roots it will have can be predicted by examining the discriminant $D = b^2 - 4ac$:

- If $D < 0$, the roots are imaginary and occur as a pair of complex conjugates.
- If $D = 0$, the roots are real, rational, and equal.
- If $D > 0$, the roots are real and unequal. If $b^2 - 4ac$ is a perfect square, the roots are rational; otherwise the roots are irrational and occur in conjugate pairs.

KEY EXAMPLE

Without solving $3x^2 - 11x = 4$, describe the nature of its roots.

Solution: The quadratic equation in standard form is $3x^2 - 11x - 4 = 0$. Because $a = 3$, $b = -11$, and $c = -4$,

$$D = b^2 - 4ac = 169.$$

Since 169 is positive and a perfect square, the roots are real, rational, and unequal.

KEY EXAMPLE

Find the smallest integer value of k such that the roots of the quadratic equation $x^2 - 5x + k = 0$ are complex.

Solution: For the equation $x^2 - 5x + k = 0$, $a = 1$, $b = -5$, and $c = k$. Then

$$D = b^2 - 4ac = 25 - 4k.$$

If the roots are complex, the discriminant is less than 0. Hence, $25 - 4k < 0$. Solving for k yields the inequality $k > 6\frac{1}{4}$. The *smallest integer* value of k that satisfies the inequality is 7. Thus, $k = 7$.

Key 29   Sum and product of the roots

OVERVIEW   *From the quadratic formula we know that the two roots are*

$$x_1 = \frac{-b+\sqrt{b^2-4ac}}{2a} \quad and \quad x_2 = \frac{-b-\sqrt{b^2-4ac}}{2a}.$$

Adding and multiplying these expressions produce convenient formulas that show the relationships between the coefficients of a quadratic equation in standard form and the sum and product of its roots.

Sum and product of the roots: If x_1 and x_2 represent the two roots of the quadratic equation $ax^2 + bx + c = 0$ $(a \neq 0)$, then:

$$\text{Sum of roots} = x_1 + x_2 = -\frac{b}{a}$$

and

$$\text{Product of roots} = x_1 \cdot x_2 = \frac{c}{a}.$$

Furthermore, dividing each term of $ax^2 + bx + c = 0$ by the coefficient of x^2 produces the equation

$$x^2 + \frac{b}{a}x + \frac{c}{a} = 0,$$

which has the general form

$$x^2 + [-(sum\ of\ roots)]\ x + (product\ of\ roots) = 0.$$

KEY EXAMPLE

If a root of the equation $x^2 + px + q$ is $2 + \sqrt{5}$, find the values of p and q.

Solution: Since irrational roots of quadratic equations occur in conjugate pairs, the other root is $2 - \sqrt{5}$. Hence, the sum of the two roots is 4 and their product is $(2 + \sqrt{5})(2 - \sqrt{5}) = 4 - 5 = -1$. Since p, the coefficient of x in the given quadratic equation, is the negative of the sum of the roots and q equals the product of the roots, $p = -4$ and $q = -1$.

KEY EXAMPLE

Write a quadratic equation whose roots are $3 + i$ and $3 - i$.

Solution: A quadratic equation whose roots are known can be formed by writing $x^2 + [-(\text{sum of roots})] x + (\text{product of roots}) = 0$. Since the sum of the given roots is 6 and their product is $(3 + i)(3 - i) = 9 - i^2 = 10$, the quadratic equation that has $3 + i$ and $3 - i$ as its roots is $x^2 - 6x + 10 = 0$.

KEY EXAMPLE

One root of the equation $2x^2 + kx - 3 = 0$ is $\dfrac{1}{2}$. Find the other root and the value of k.

Solution: In the equation $2x^2 + kx - 3 = 0$, $a = 2$, $b = k$, and $c = -3$. The product of the roots is $\dfrac{c}{a} = \dfrac{-3}{2}$ and the given root is $\dfrac{1}{2}$. The other root must be -3 since the product of -3 and $\dfrac{1}{2}$ is $\dfrac{-3}{2}$. Substitute -3 for x in the original equation and then solve for k.

$$2(-3)^2 + k(-3) - 3 = 0$$

$$18 - 3k - 3 = 0$$

$$-3k = -15$$

$$k = \frac{-15}{-3} = 5$$

Thus, the second root is -3 and the value of k is 5.

Key 30   Graphing quadratic equations

in two variables

OVERVIEW   *The graph of the quadratic equation* y =
ax^2 + bx + c (a ≠ 0) *is a curve called a **parabola** that has a
vertical axis of symmetry. The point at which the axis of
symmetry intersects the parabola is called the **vertex** or
turning point.*

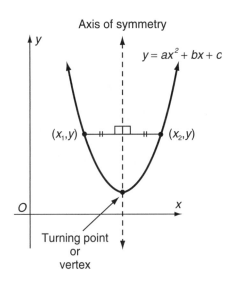

Axis of symmetry

$y = ax^2 + bx + c$

(x_1, y) (x_2, y)

O x

Turning point
or
vertex

Vertical parabola: The graph of $y = ax^2 + bx + c$ $(a \neq 0)$ is a parabola
with a vertical axis of symmetry. An equation of this axis of sym-
metry is $x = \dfrac{-b}{2a}$. Thus, the x-coordinate of the turning point is $\dfrac{-b}{2a}$.

Substituting this value for x in $y = ax^2 + bx + c$ gives the y-coordinate
of the turning point. (See Figure 5.1.) The sign of a, the coefficient
of x^2, determines whether the parabola opens "up" or "down."
Observe in Figure 5.1 the following:

• If $a > 0$, the parabola opens up (can "hold water") and the turning
  point is a *minimum* point on the parabola.
• If $a < 0$, the parabola opens down ("spills water") and the turning
  point is a *maximum* point on the parabola.

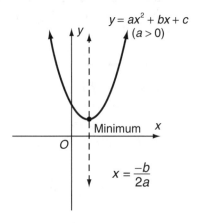

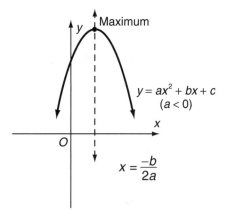

Figure 5.1 Effect of the Sign of the Coefficient of x^2 on the Parabola

Graph of $y = ax^2$ ($a \neq 0$): The parabola $y = ax^2$ has the origin as its turning point and the y-axis as its axis of symmetry. (See Figure 5.2.)

- As $|a|$ increases, the parabola becomes narrower horizontally. Thus, the graph of $y = 2x^2$ is narrower than the graph of $y = x^2$.
- As $|a|$ decreases, the parabola becomes broader horizontally.

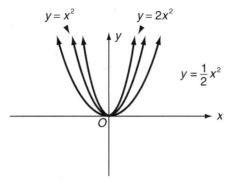

Figure 5.2   Graph of $y = ax^2$

Graph of $y = ax^2 + c$ ($a \neq 0$): The parabola $y = ax^2 + c$ is the graph of $y = ax^2$ shifted vertically $|c|$ units. (See Figure 5.3.)

- If $c > 0$, the shift is up.
- If $c < 0$, the shift is down.

Figure 5.3 illustrates the effect of c for the cases in which $c = 2$ and $c = -2$.

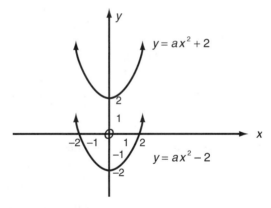

Figure 5.3   Graph of $y = ax^2 + c$

Graphing a parabola with a calculator: To use a graphing calculator to graph the parabola $y = 2x^2 - 8x + 7$, set $Y_1 = 2X \wedge 2 - 8X + 7$, where $X \wedge 2$ is entered in the $Y =$ editor by pressing either

$$\boxed{\text{x,T,}\theta\text{,n}} \; \boxed{x^2} \quad \text{or} \quad \boxed{\text{x,T,}\theta\text{,n}} \; \boxed{\wedge} \; \boxed{2}.$$

When using a calculator to display the graph of a parabola, you may have to adjust the window size in order to be able to view the key features of the graph.

Using a calculator to find the vertex of a parabola: The **minimum** or **maximum** feature in the CALC menu of your calculator can be used to find the coordinates of the vertex of a parabola. For example, to find the coordinates of the vertex of $Y_1 = 2X \wedge 2 - 8X + 7$, display the graph in an appropriate window, such as $[-4.7, 4.7] \times [-3.1, 3.1]$. Then proceed as follows:

- Open the CALC menu by pressing $\boxed{\text{2nd}}$ $\boxed{\text{CALC}}$. Since the vertex of $y = 2x^2 - 8x + 7$ is the lowest point on the parabola, select 3:**minimum**. (If the vertex is the highest point on the parabola, select 4:**maximum**.)

- Move the cursor slightly to the left of the vertex, and press $\boxed{\text{ENTER}}$.

  Move the cursor slightly to the right of the vertex, and press $\boxed{\text{ENTER}}$ two times.

- Read the coordinates of the vertex at the bottom of the window, as shown in Figure 5.4. If you do not get the exact value for the x-coordinate, you may need to adjust the window variables so that you are looking at the graph in a friendly window.

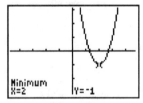

Figure 5.4 Locating the Vertex

Figure 5.5 Table of Values

Graphing a parabola using a table of values: If you need to draw the parabola $y = ax^2 + bx + c$ on graph paper, find the x-coordinate of the vertex by using the formula $x = \dfrac{-b}{2a}$. Next, plot three points on either side of the vertex, and connect these points with a smooth curve. You can find these three points by using the built-in table feature of your graphing calculator. For example, if the equation $y = 2x^2 - 8x + 7$ has already been stored in your calculator, you can obtain the table of values in Figure 5.5 as follows:

- Press $\boxed{\text{2nd}}$ $\boxed{\text{TBLSET}}$.
- Change the **TblStart** value to –1 since –1 is 3 units less than the x-coordinate of the vertex. If necessary, set ΔTbl = 1 so that x increases in steps of 1 unit.
- Press $\boxed{\text{2nd}}$ $\boxed{\text{TABLE}}$. If you need to look at table entries that are not currently in view, use a cursor key to scroll up or down.

You can easily tell from the table that the coordinates of the vertex are (2, –1), as corresponding points above and below this point have the same y-coordinate.

Making graphical connections: At an x-intercept of a graph, $y = 0$. Thus, if the graph of $y = ax^2 + bx + c$ has x-intercepts, the x-coordinates of those points correspond to the real roots of $ax^2 + bx + c = 0$. The accompanying table summarizes the connections between the discriminant and the graph of a quadratic equation.

Relating the Discriminant to the Graph of $y = ax^2 + bx + c$ ($a \neq 0$)

Discriminant	Type of Roots	Graph
$b^2 - 4ac > 0$	Real and unequal roots. If $b^2 - 4ac$ is a perfect square, the roots will be rational.	Two x-intercepts.
$b^2 - 4ac = 0$	Two equal rational roots.	One x-intercept at vertex
$b^2 - 4ac < 0$	Two imaginary roots.	No x-intercept

Horizontal parabola: The graph of $x = ay^2 + by + c$ $(a \neq 0)$ is a parabola with a horizontal axis of symmetry. An equation of this axis of symmetry is $y = \dfrac{-b}{2a}$. As illustrated in Figure 5.6, if $a > 0$, the parabola opens to the right; if $a < 0$, the parabola opens to the left.

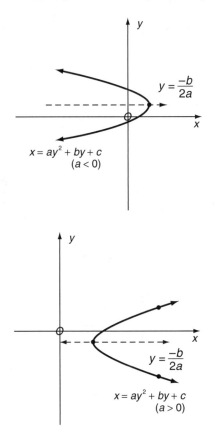

Figure 5.6 Parabolas with Horizontal Axes of Symmetry

Key 31   Solving quadratic equations graphically

OVERVIEW   *A graphing calculator can be used to solve quadratic equations with real roots.*

Solving a quadratic equation graphically: To solve a quadratic equation such as $x^2 - 2x = 3$ graphically, set $Y1 = x^2 - 2x$ and $Y2 = 3$. The solution is the point or points at which the graphs of $Y1$ and $Y2$ intersect. To find the points of intersection, use the **intersect** feature from the CALC menu of your calculator, as shown in Figure 5.7. The solutions are $x = -1$ and $x = 3$.

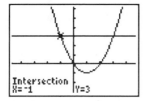 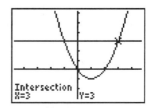

Figure 5.7 Solving $x^2 - 2x = 3$ Graphically

KEY EXAMPLE

Solve $x^2 - 3x - 2 = 0$ graphically, and approximate the roots to the *nearest hundredth*.

Solution: Set $Y_1 = X ^\wedge 2 - 3X - 2$, and display its graph in a friendly window, such as $[-4.7, 4.7] \times [-6.2, 6.2]$, as shown in the accompanying figure. Since $x^2 - 3x - 2$ is not factorable, the *x*-intercepts of the parabola represent irrational roots of $x^2 - 3x - 2 = 0$. From the graph on the left, you know that one root is between 0 and -1, and the other root is between 3 and 4.

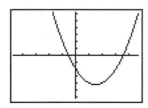 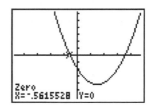

To approximate the negative root, select **2: zero** from the CALC menu.

- Move the cursor to a point on the graph that is slightly to the left of the negative x-intercept. Press ENTER.
- Move the cursor to a point on the graph that is slightly to the right of the negative x-intercept. Press ENTER two times.
- Read from the bottom left corner of the viewing window that the x-intercept is –0.5615528, as shown in the graph on the right.

By moving the cursor slightly to the left of the positive x-intercept and following the same procedure, you can verify that the positive x-intercept is 3.5615528. Hence, the roots of $x^2 - 3x - 2 = 0$, to the *nearest hundredth*, are **–0.56** and **3.56**.

Solving a linear-quadratic system of equations: A linear-quadratic system of equations can be solved either graphically or algebraically.

- To solve the system $y = -x^2 + 4x - 3$ and $x + y = 1$ graphically using a calculator, graph the parabola and a line in a friendly window, such as $[-4.7, 4.7] \times [-6.2, 6.2]$. Then use the **intersect** feature to determine the coordinates of the points of intersection, as shown in Figure 5.8.

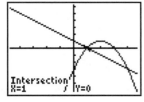 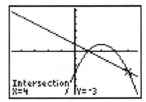

Figure 5.8 Using the Intersect Feature in the CALC Menu to Find the Points of Intersection of $y = -x^2 + 4x - 3$ and $x + y = 1$

- To solve the system $y = -x^2 + 4x - 3$ and $x + y = 1$ algebraically, solve the linear equation for y and then replace y with that expression in the quadratic equation. Since $x + y = 1$,

$$y = 1 - x, \text{ so } y = 1 - x = -x^2 + 4x - 3,$$

which, after simplifying, becomes $x^2 - 5x + 4 = 0$. Thus,

$$(x - 1)(x - 4) = 0, \text{ so } x = 1 \text{ or } x = 4.$$

(i) If $x = 1$, then $1 + y = 1$, so $y = 0$. Thus, $(1, 0)$ is a solution.
(ii) If $x = 4$, then $4 + y = 1$, so $y = -3$. Hence, $(4, -3)$ is a solution.

Theme 6 FUNCTIONS AND
THEIR GRAPHS

*F*unctions are powerful mathematical tools for describing how two variables are related to each other. A function may take many forms. Regardless of the form it takes, a function must satisfy the following two conditions:

- There is a "rule" that tells how each possible value of one variable is matched with a value of the other variable. This rule may be specified numerically, algebraically, graphically, or verbally.
- No two ordered pairs of the form (x, y) have the same x-value but different y-values.

Key 32 Functions

OVERVIEW *A function may take the form of a set of ordered pairs, a graph, or an equation. Regardless of the form it takes, a function must obey the condition that no two of its ordered pairs have the same first member with different second members.*

Relation: A set of ordered pairs of the form (x, y) is called a relation.

Function: A relation in which no two ordered pairs have the same x-value but different y-values is called a function. Functions are usually named by lower-case letters such as f, g, and h. For example, if $f = \{(-3, 9), (0, 0), (3, 9)\}$ and $g = \{(4, -2), (4, 2)\}$ then f is a function but g is *not* a function.

Domain and range: The domain of a function is the set of all the first members of its ordered pairs, and the range of a function is the set of all the second members of its ordered pairs. If function $f = \{(a, A), (b, B), (c, C)\}$, then its *domain* is $\{a, b, c\}$ and its *range* is $\{A, B, C\}$.

Functions as mappings: A function may be viewed as a *mapping* or a pairing of the elements of one set with the elements of a second set in such a way that each element of the first set, the domain, is paired with exactly one element of the second set, called the **codomain**. For example, if a function f maps $\{a, b, c\}$ into $\{A, B, C, D\}$ so that $a \rightarrow A$ (read as "a is mapped onto A"), $b \rightarrow B$, and $c \rightarrow C$, then the domain is $\{a, b, c\}$ and the codomain is $\{A, B, C, D\}$. Since a is paired with A in the codomain, A is called the **image** of a. Each element of the codomain that corresponds to an element of the domain is the image of that element. The set of image points, $\{A, B, C\}$, is called the **range**. Thus, the range is a subset of the codomain.

Onto mappings: Set A is mapped *onto* set B if each element of set B is the image of an element of set A. Thus, every function maps its domain *onto* its range.

Describing a function by an equation: The rule by which each x-value gets paired with the corresponding y-value may be specified by an equation. For example, the function described by the equation $y = x + 1$ requires that for any choice of x in the domain, the corresponding range value is $x + 1$. Thus, $2 \rightarrow 3$, $3 \rightarrow 4$, and $4 \rightarrow 5$.

Restricting domains of functions: Unless otherwise indicated, the domain of a function is assumed to be the largest possible set of real numbers. Thus, the domain of $y = \dfrac{x}{x^2 - 4}$ is the set of all real numbers *except* ± 2 since for these values of x the denominator is 0. The domain of $y = \sqrt{x-1}$ is the set of real numbers greater than or equal to 1 since, for any value of x less than 1, the square root radical has a negative radicand; thus, the radical does not represent a real number.

KEY EXAMPLE

Determine which of the following relations, if any, describes functions:

(a) $y = \sqrt{x}$ (b) $y = x^3$ (c) $y > x$ (d) $x = y^2$

Solution: Equations (a) and (b) produce exactly one value of y for each value of x. Hence, equations (a) and (b) describe functions. In (c), $y > x$ does *not* represent a function since it contains ordered pairs such as (1, 2) and (1, 3) in which the same value of x is paired with different values of y. In (d), the equation $x = y^2$ is *not* a function since ordered pairs such as (4, 2) and (4, –2) satisfy the equation but have the same value of x paired with different values of y.

Function notation: For any function f, the value of y that corresponds to a given value of x is denoted by $f(x)$. If $y = 5x - 1$, then $f(2)$, read as "f of 2," represents the value of y when $x = 2$. Thus,

$$f(2) = 5 \cdot 2 - 1 = 9.$$

If $x = 3$, then

$$f(3) = 5 \cdot 3 - 1 = 14.$$

In an equation that describes function f, $f(x)$ may be used in place of y, as in $f(x) = 5x - 1$. If $y = f(x)$, then y is said to be a function of x. Since the value of y depends on the value of x, y is called the **dependent variable** and x is the **independent variable**.

Using variable names other than x and y: Functions may be expressed in terms of variables other than x and y. For example, the area, A, of a circle depends on the radius, r, of the circle. If the name of this function is g, then $A = g(r)$. For function g, the independent variable is r and the dependent variable is A. If a function h is described by the equation $p = -q^2 + 23q + 47$, then p is a function of q. Hence, $p = h(q)$, where q is the independent variable and p is the dependent variable.

KEY EXAMPLES

1. If $f(x) = (x + 1)^{3/2}$, find $f(8)$.

Solution: $\qquad f(8) = (8+1)^{3/2} = (\sqrt{9})^3 = 27$

2. If $g(t) = t^2$, find $g(x + 1) - g(x)$.

Solution: Since

$$g(x) = (x + 1)^2 = x^2 + 2x + 1 \text{ and } g(x) = x^2,$$

$$g(x + 1) - g(x) = (x^2 + 2x + 1) - (x^2) = 2x + 1.$$

3. If $h = \{(-2, 5), (-1, 2), (0, 1), (1, 2), (2, 5)\}$, find $h(1)$.

Solution: The function value $h(1)$ represents the y-value in h, the ordered pair whose x-value is 1. Since h contains $(1, 2)$, $h(1) = 2$.

Forming new functions using arithmetic operations: Two functions may be added, subtracted, multiplied, or divided to produce a third function whose domain is the set of values that are common to the domains of the original two functions.

Example: If $f(x) = x^2$ and $g(x) = x - 1$, then

$$f(x) + g(x) = x^2 + x - 1,$$

$$f(x) - g(x) = x^2 - x + 1,$$

$$f(x) \cdot g(x) = x^3 - x^2,$$

and

$$\frac{f(x)}{g(x)} = \frac{x^2}{x-1},$$

provided $x \neq 1$.

Key 33 Vertical and horizontal line tests

OVERVIEW *There are simple tests for telling whether a graph is a function and, if it is a function, whether the same value of* y *is used in different ordered pairs.*

Vertical line test: A graph represents a function only if no vertical line can be drawn that intersects the graph in more than one point. See Figures 6.1 and 6.2.

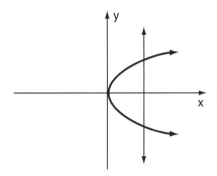

Figure 6.1 Graph Is Not a Function

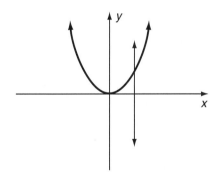

Figure 6.2 Graph Is a Function

A function is a **one-to-one** function if different values of x are never paired with the same value of y.

Horizontal line test: Thus, a function is a one-to-one function if no horizontal line intersects its graph in more than one point. The graph in Figure 6.2 fails the horizontal line test. The graph in Figure 6.3 passes the horizontal line test.

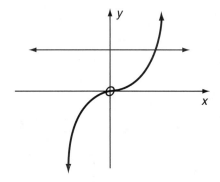

Figure 6.3 Graph Is a One-to-One Function

Key 34 Symmetry and functions

OVERVIEW   *A graph is* symmetric *with respect to a line if the line divides the figure into two parts so that, when the coordinate plane is "folded" along the line, the two parts of the graph coincide. A graph may have one of the coordinate axes as a line of symmetry.*

Definition and test of symmetry with respect to a coordinate axis:
Figures 6.4 and 6.5 show graphs with y-axis and x-axis symmetry, respectively. The graph of an equation has y-axis or x-axis symmetry if the following conditions are satisfied:

- **y-Axis symmetry** exists if whenever (x, y) is on the graph, $(-x, y)$ is also on the graph. To test for y-axis symmetry without graphing an equation, replace x with $-x$. If this produces an equivalent equation, then the graph of this equation is symmetric with respect to the y-axis.

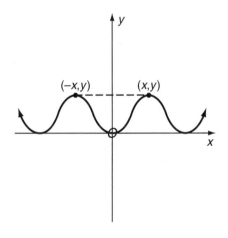

Figure 6.4 y-Axis Symmetry

Example: Replacing x with $-x$ in the equation $y = x^2 + 1$ gives $y = (-x)^2 + 1 = x^2 + 1$. Since this substitution does not change the original equation, the graph of $y = x^2 + 1$ is symmetric with respect to the y-axis.

- **x-Axis symmetry** exists if whenever (x, y) is on the graph, $(x, -y)$ is also on the graph. To test for x-axis symmetry without graphing an equation, replace y with $-y$. If this produces an equivalent equation, then the graph of this equation is symmetric with respect to the x-axis.

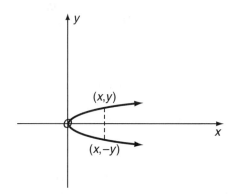

Figure 6.5 *x*-Axis Symmetry

Example 1: Replacing y with $-y$ in the equation $x = y^2 - 3$ gives $x = (-y)^2 - 3 = y^2 - 3$. Since this substitution does not change the original equation, the graph of $x = y^2 - 3$ is symmetric with respect to the x-axis.

Example 2: The circle whose equation is $x^2 + y^2 = 16$ has both x-axis and y-axis symmetry since $(-x)^2 + (-y)^2 = x^2 + y^2 = 16$.

Definition and test for origin symmetry: A graph has origin symmetry if whenever (x, y) is on the graph, $(-x, -y)$ is also on the graph. The graph in Figure 6.6 has origin symmetry. To test for origin symmetry without graphing an equation, replace x with $-x$ and y with $-y$. If this produces an equivalent equation, then the graph of this equation is symmetric with respect to the origin.

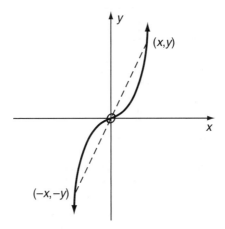

Figure 6.6   Origin Symmetry

Example: The graph of $y = \dfrac{8}{x}$ $(x \neq 0)$ has origin symmetry since

replacing x and y with their opposites gives $-y = \dfrac{8}{-x}$, which, after

multiplying both sides of the equation by -1, can be written as

$y = \dfrac{8}{x}$.

Even function: An even function has the property that $f(-x) = f(x)$ for all x in the domain of f. Thus, the graph of an even function has y-axis symmetry.

Example: The function $f(x) = x^4 - x^2 + 3$ is an even function since

$$f(-x) = (-x)^4 - (-x)^2 + 3 = x^4 - x^2 + 3.$$

Odd function: An odd function has the property that $f(-x) = -f(x)$ for all x in the domain of f. Thus, the graph of an odd function has origin symmetry.

Example 1:   The function $f(x) = x^3 - x$ is an odd function since

$$f(-x) = (-x)^3 - (-x) = -x^3 + x = -(x^3 - x) = -f(x).$$

Example 2:   The functions $f(x) = x^2 + 2x$ and $f(x) = x^3 + 1$ are neither odd nor even.

Key 35 Graphs of special functions

OVERVIEW *Some types of functions occur so frequently in mathematics that you should know their graphs.*

Reciprocal function: The graph of the reciprocal function $f(x) = \dfrac{1}{x}$ (or $xy = 1$) is called an **equilateral (rectangular) hyperbola** (see Figure 6.7). An equilateral hyperbola consists of two branches that are located in Quadrants I and III. The domain and range consist of all real numbers except $x, y \neq 0$. The coordinate axes are **asymptotes** of the curve since each branch gets closer and closer to a coordinate axis, but never touches it. The function is *odd* so its graph is symmetric with respect to the origin. In general, the graph of $f(x) = \dfrac{k}{x}$ (or $xy = k$), where k is some nonzero constant, is an equilateral hyperbola. The sign of k determines the pair of opposite quadrants in which the two branches are located. If $k > 0$, then the branches are in Quadrants I and III; if $k < 0$, the branches are in Quadrants II and IV.

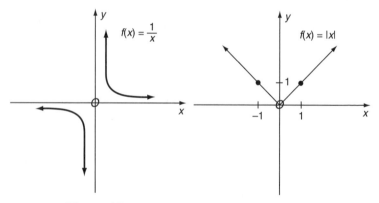

Figure 6.7

Graph of $f(x) = \dfrac{1}{x}$

Figure 6.8

Graph of $f(x) = |x|$

Absolute value function: The domain of the absolute value function $f(x) = |x|$ is the set of real numbers, and its range is the set of non-negative real numbers. The graph, as shown in Figure 6.8, consists of the rays $y = x$ and $y = -x$, whose common endpoint is the origin. The function is *even* so its graph is symmetric with respect to the y-axis.

Square root function: The domain and range of the square root function $f(x) = \sqrt{x}$ are limited to the set of real nonnegative numbers. In Figure 6.9, the graph starts at the origin and gradually rises in Quadrant I.

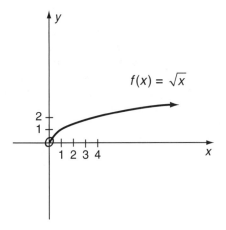

Figure 6.9   Graph of $f(x) = \sqrt{x}$

Cube function: The domain and range of the cube function $f(x) = x^3$ is the set of real numbers. The graph, as shown in Figure 6.10, is steeper than the graph of $f(x) = x^2$ and is restricted to Quadrants I and III. Since the function is *odd*, the graph is symmetric with respect to the origin.

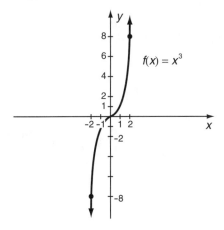

Figure 6.10   Graph of $f(x) = x^3$

Greatest integer [step] function: This function is denoted by **[x]**, read as "the greatest integer less than or equal to x," and is defined by the equation

$$f(x) = [x] = n$$

where n is an integer such that $n \le x < n + 1$. For example, $[3.9] = 3$ since $3 \le 3.9 < 4$; that is, 3 is the greatest integer ≤ 3.9. To illustrate further, $[0.8] = 0$ because $0 \le 0.8 < 1$. Also, $[-1.7] = -2$ since $-2 \le -1.7 < -1$. Figure 6.11 shows the graph of $y = [x]$ on $-2 \le x < 4$. Since the range of this function is restricted to integer values while the domain consists of real numbers, the graph shown in Figure 6.11 is discontinuous and consists of a rising set of horizontal "steps."

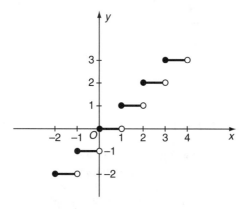

Figure 6.11   Graph of $y = [x]$

Piecewise defined functions: The definitions of piecewise defined functions change over different intervals on their domains. The graph of

$$f(x) = \begin{cases} x^2 & \text{if } x < 0 \\ \sqrt{x} & \text{if } x \ge 0 \end{cases}$$

is shown in Figure 6.12. If $x < 0$, then $f(x)$ is defined as x^2, and for $x \le 0$, $f(x)$ is defined as $\sqrt{x}$. Thus, $f(-4) = (-4)^2 = 16$ and $f(4) = \sqrt{4} = 2$.

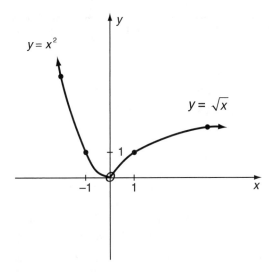

Figure 6.12 Graph of a Piecewise Defined Function

Example: The graph of

$$g(x) = \begin{cases} -x^2 + 1, x \le 1 \\ 3x, x > 1 \end{cases}$$

is shown. The graph is not continuous since it has a "jump" discontinuity at the "crossover point" at $x = 1$.

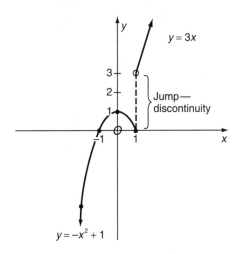

Key 36 Shifts of graphs of functions

OVERVIEW *Two functions may be related so that the graph of one function can be obtained by shifting the graph of a related but simpler function.*

Vertical shift: If k is a constant and $g(x) = f(x) + k$, then the graph of $g(x)$ is the graph of $f(x)$ shifted *vertically* $|k|$ units. The graph of $g(x)$ is shifted *up* if $k > 0$, and is shifted *down* if $k < 0$. For example, the graph of $g(x) = x^2 + 3$ can be obtained by shifting the graph of $f(x) = x^2$ vertically up 3 units so its turning point is shifted from $(0, 0)$ to $(0, 0 + 3) = (0, 3)$. The graph of $g(x) = x^2 - 1$, is the parabola $y = x^2$ shifted vertically down 1 unit so its turning point is $(0, -1)$.

Horizontal shift: If h is a constant and $g(x) = f(x - h)$, then the graph of $g(x)$ is the graph of $f(x)$ shifted *horizontally* $|h|$ units. The graph of $f(x)$ is shifted to the right if $h > 0$, and is shifted to the left if $h < 0$.

KEY EXAMPLE

Sketch the graphs of: (a) $f(x) = \sqrt{x-1}$ (b) $f(x) = \sqrt{x} - 1$

Solution: (a)   The graph of $f(x) = \sqrt{x-1}$ is obtained by shifting the graph of $y = \sqrt{x}$ 1 unit horizontally to the right.

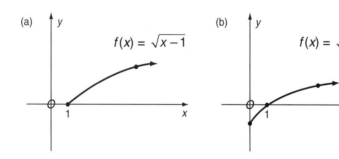

(b)   The graph of $f(x) = \sqrt{x} - 1$ is obtained by shifting the graph of $y = \sqrt{x}$ 1 unit vertically down.

Graphing $y = a(x - h)^2 + k$:   Some graphs are combinations of horizontal and vertical shifts of familiar graphs. For example, if $g(x) = (x - 2)^2 + 1$, then $h = 2$ and $k = 1$. Hence, the graph of $g(x)$ can be obtained by shifting the graph of $y = x^2$ horizontally 2 units to the right and vertically 1 unit up. Since the turning point of $y = x^2$ is $(0, 0)$ the turning point of $g(x) = (x - 2)^2 + 1$ is $(0 + 2, 0 + 1) = (2, 1)$. In general, the graph of $y = a(x - h)^2 + k$ is the parabola $y = ax^2$ shifted so that its turning point is (h, k) and its axis of symmetry is $x = h$.

KEY EXAMPLE

Sketch the graph of $f(x) = |x + 2| + 1$

Solution: Since

$$f(x) = |x + 2| + 1 = |x - (-2)| + 1,$$

$h = -2$ and $k = 1$. Hence, the graph of $f(x)$ is obtained by shifting the graph of $y = |x|$ 2 units horizontally to the left and 1 unit vertically up.

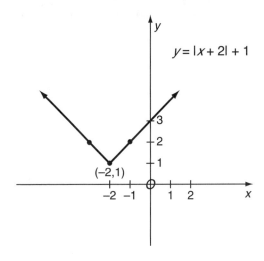

KEY EXAMPLE

Put the equation $y = 3x^2 + 6x - 5$ into the form $y = a(x - h)^2 + k$ and describe its graph.

Solution: After writing $y = 3(x^2 + 2x) - 5$, complete the square for the expression inside the parentheses. The number that completes the square is obtained by taking one-half the coefficient of x and squaring it, which results in 1. Since this number is multiplied by 3, at the same time subtract 3 times this value outside the parentheses:

$$y = 3(x^2 + 2x + 1) - 3 \cdot 1 - 5 = 3(x + 1)^2 - 8$$

Since

$$y = 3(x + 1)^2 - 8$$

is equivalent to

$$y = 3[x - (-1)]^2 + (-8),$$

$h = -1$ and $k = -8$. Thus, the graph of $y = 3(x + 1)^2 - 8$ is the parabola $y = 3x^2$ shifted 1 unit horizontally to the left and 8 units vertically down so that its new turning point is $(-1, -8)$ and its axis of symmetry is $x = -1$.

Key 37 The composition of two
functions

OVERVIEW   *Starting with two functions* f *and* g, *we can define a new function,* h, *by taking a function of a function and writing* h(x) = g(f(x)). *Function* h *is called a* **composite function**.

Example of a composite function: If $f(x) = x - 5$ and $g(x) = \sqrt{x}$, then $g(f(9))$ represents the function value that results when $x = 9$ is the input value for function f, and the corresponding range value, $f(9)$, becomes the input value of function g. Since $f(9) = 9 - 5 = 4$ and $g(4) = \sqrt{4} = 2$, $g(f(9)) = g(4) = 2$.

Composition of functions f **and** g: The **composite of function** f **followed by function** g is denoted by $g \circ f$ (read as "g of f") and is defined as the set of function values

$$(g \circ f)(x) = g(f(x))$$

for all x that is in the domain of f such that $f(x)$ is in the domain of g. The notation $g(f(x))$ represents a particular function value and is read as "g of f of x."

Order of composition matters: Functions f and g can be composed in either order. The **composite of function** g **followed by function** f is denoted by $f \circ g$ and consists of the set of function values $f(g(x))$. In the first example of a composite function, where $f(x) = x - 5$ and $g(x) = \sqrt{x}$, we saw that $g(f(9)) = 2$. Similarly, $f(g(9)) = f(3) = -2$. Thus, $g(f(9)) \neq f(g(9))$. In general, the composition of functions is *not* a commutative operation.

Domain of a composite function: A composite function, for example $g \circ f$, is defined so that the domain of $g \circ f$ is restricted to the largest possible set of real numbers for which x is in the domain of function f *and* $f(x)$ is in the domain of function g. For example, if $f(x) = x - 5$ and $g(x) = \sqrt{x}$ $(x \geq 0)$, then the domain of the composite function $g \circ f$ consists only of those values of x in the domain of function f for which $g(f(x))$ is real. If x is less than 5, then $f(x)$ is negative, so $g(f(x))$ is not real. Thus, if $x = 4$, then $f(4) = -1$ and $g(f(4)) = \sqrt{-1}$, so $g(f(4))$ is not real. Hence, the domain of $g \circ f$ must be restricted to all numbers x that are greater than or equal to 5.

Finding an equation for a composite function: If $f(x) = x^2$ and $g(x) = 3x + 1$, then an equation of the composite function $g \circ f$ can be obtained by starting with $g(f(x))$ and replacing $f(x)$ with x^2, and then applying the function rule for $g(x)$. Thus,

$$g(f(x)) = g(x^2) = 3[x^2] + 1.$$

Similarly, the composite function $f \circ g$ can be derived by starting with $f(g(x))$ and replacing $g(x)$ with $3x + 1$, and then applying the function rule for $f(x)$:

$$f(g(x)) = f(3x + 1) = [3x + 1]^2 = 9x^2 + 6x + 1.$$

Since $g(f(x))$ and $f(g(x))$ are defined by different equations, once again we see that the composition of two functions is *not* a commutative operation.

KEY EXAMPLE

If $f = \{(1, 3)(2, 4)\}$ and $g = \{(1, 2)(3, 6)\}$, find

$\qquad$ (a) $(g \circ f)(1)$ $\qquad$ (b) $(f \circ g)(1)$ $\qquad$ (c) $g(f(2))$

Solution: (a) $(g \circ f)(1) = g(f(1)) = g(3) = 6$

(b) $(f \circ g)(1) = f(g(1)) = f(2) = 4$

(c) $g(f(2)) = g(4)$. Since 4 is not in the domain of g, $g(f(2))$ is not defined.

KEY EXAMPLE

If $g(x) = \dfrac{x-1}{2}$ and $h(x) = 2x + 1$, find: (a) $h(g(x))$ (b) $h(h(3))$

Solutions: (a) $h(g(x)) = h\left(\dfrac{x-1}{2}\right) = 2\left(\dfrac{x-1}{2}\right) + 1 = x$

(b) Since $h(h(x)) = h(2x) = 2(2x) = 4x$, $h(h(3)) = 4 \cdot 3 = 12$.

KEY EXAMPLE

If $g(x) = \sqrt{x-2}$ and $h(x) = \left(\dfrac{1}{x}\right)^2$, find $h(g(x))$ and state the domains of functions g and $h \circ g$.

Solution: For $h(g(x)) = \dfrac{1}{x-2}$, $x \neq 2$. Since the domain of $g(x)$ is $x \geq 2$ and $x \neq 2$, the domain of $h \circ g$ is $x > 2$.

Key 38 Inverse functions

OVERVIEW *For some functions there exist other functions that can reverse their effects. Pairs of functions that are related in this way are called **inverse functions.***

Example of inverse functions: Functions such as $f(x) = x^2$ and $g(x) = \sqrt{x}$ are inverse functions since each undoes the effect of the other. For example, if $x = 3$, then $f(3) = 9$ and $g(9) = \sqrt{9} = 3$ which is the value of x we started with. Thus, $g(f(3)) = 3$. Also, $f(g(3)) = 3$. Notice that the ordered pair $(3, 9)$ belongs to function f and the corresponding ordered pair for function g is $(9, 3)$. If functions f and g are inverse functions, then the domain of f is the range of g, and the range of f is the domain of g.

Definition of inverse functions: Functions f and g are inverse functions if their compositions, taken in either order, always produce the original input value x. To show that functions f and g are inverse functions, verify that

$$g(f(x)) = f(g(x)) = x$$

for all values for which the compositions are defined.

Forming inverse functions: To find the inverse of a function, if it exists, we need to interchange its domain and range.

- If the function is defined by a set of ordered pairs, then its inverse is obtained by swapping the positions of the first and second members of each ordered pair. For example, if $f = \{(-1, 0), (0, 1), (1, 2)\}$, then the inverse of f is $\{(0, -1), (1, 0), (2, 1)\}$.
- If the function is defined by an equation, then its inverse is formed by interchanging x and y, and then solving the resulting equation for y. If $f(x) = 2x + 3$, let $y = 2x + 3$. Interchanging x and y yields $x = 2y + 3$. Solving this equation for y gives the inverse of $f(x)$,
$$y = \frac{x-3}{2}.$$

Inverse functions: Not all functions have inverse functions. Interchanging the x and y values of the function $f = \{(-2, 4), (2, 4)\}$ gives $\{(4, 2), (4, -2)\}$, which is not a function. If f is a one-to-one function (f passes the horizontal line test), then its inverse is a function and is denoted by f^{-1}. Be careful not to interpret the -1 in f^{-1} as an exponent. The notation f^{-1} is read as "f–inverse."

KEY EXAMPLE

Decide whether $f(x) = x^3 - 1$ and $g(x) = (x + 1)^{\frac{1}{3}}$ are inverse functions.

Solution: Determine whether the statements $f(g(x)) = x$ and $g(f(x)) = x$ are both true:

$$f(g(x)) = f\left((x+1)^{\frac{1}{3}}\right) \qquad\qquad g(f(x)) = g(x^3 - 1)$$

$$= \left((x+1)^{\frac{1}{3}}\right)^3 - 1 \quad \text{and} \quad = ((x^3 - 1) + 1)^{\frac{1}{3}}$$

$$= (x + 1) - 1 \qquad\qquad = (x^3)^{\frac{1}{3}}$$

$$= x \qquad\qquad\qquad = x$$

Since both statements are true, functions f and g are inverse functions.

Making functions one-to-one: Sometimes the domain of a function must be restricted so that its inverse can be obtained. For example, $f(x) = x^2$ is not a one-to-one function. By restricting the domain of f to the set of nonnegative real numbers ($x \geq 0$), we insure that f has an inverse, namely, $f^{-1}(x) = \sqrt{x}$.

Graphs of inverse functions: For every point (a, b) contained on the graph of a one-to-one function, point (b, a) is contained on the graph of its inverse. As shown in Figure 6.13, this means that the graph of a function and its inverse are symmetric with respect to the line $y = x$. Either graph may be obtained by reflecting the other graph in the line $y = x$.

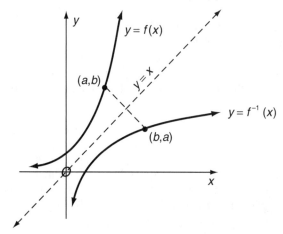

Figure 6.13 Symmetry of Graphs of Inverse Functions

Key 39 Equations that describe variation

OVERVIEW *Sometimes two or more variables are related by the operation of multiplication or division or by a combination of these operations.*

Direct variation: If variables x and y are related so that the ratio of y to x always remains the same, we say that y *varies directly as x*. This relationship is expressed in symbols as

$$\frac{y}{x} = k \quad \text{or} \quad y = kx$$

where k stands for a number called the **constant of variation**. Multiplying x or y by a fixed amount always causes the other variable to be *multiplied* by the same fixed amount.

Inverse variation: If variables x and y are related so that their product always remains the same, we say that y *varies inversely as* x. This relationship is expressed in symbols as

$$xy = k \quad \text{or} \quad y = \frac{k}{x}$$

where k is a nonzero number called the **constant of variation**. Multiplying x or y by some fixed amount always causes the other variable to be *divided* by the same fixed amount. The graph of an inverse variation is an equilateral hyperbola.

Variation involving powers: A variable y may vary directly or indirectly as the power of a variable x, which is illustrated in the next Key Example.

KEY EXAMPLE

If p varies directly as the square of q, and $p = 147$ when $q = 7$, find the value of q when $p = 75$.

Solution: Since p varies directly as the square of q, $p = kq^2$. Solve for k by letting $p = 147$ and $q = 7$. Thus,

$$147 = k(7^2), \text{ so } k = \frac{147}{49} = 3.$$

Using $p = 3q^2$ as the equation of variation, let $p = 75$ and solve for q, thus, $75 = 3q^2$, so $q^2 = 25$ and $q = \pm 5$.

Theme 7 POLYNOMIAL AND RATIONAL FUNCTIONS

*T*he degree of the polynomial equation $P(x) = 0$ is the degree of the polynomial $P(x)$. Some polynomial equations of degree greater than 2 can be solved by factoring. For example:

$$3x^3 + x^2 - 12x - 4 = 0$$

$$x^2(3x + 1) - 4(3x + 1) = 0$$

$$(x^2 - 4)(3x + 1) = 0$$

If $x^2 - 4 = 0$, then $x = \pm 2$. If $3x + 1 = 0$, then $x = -\frac{1}{3}$.

Other polynomial equations cannot be factored as easily or cannot be factored at all. If some of the roots of a polynomial equation of degree greater than 2 are known, then it may be possible to use this information to reduce the degree of the original equation so that a quadratic equation eventually results.

If $P(x) = 0$ is an nth degree polynomial equation, then $y = P(x)$ is a polynomial function of degree n. The graphs of polynomial functions and their quotients, called **rational functions**, can help us understand how these functions behave.

INDIVIDUAL KEYS IN THIS THEME

Key 40 Synthetic division

OVERVIEW *A condensed form of the long-division process, called* **synthetic division**, *can be used whenever a polynomial function* P(x) *is divided by a first-degree binomial of the form* x − r, *where* r *is some constant.*

Polynomial function of degree *n*: A polynomial function $P(x)$ of degree *n* has the general form

$$P(x) = a_nx^n + a_{n-1}x^{n-1} + a_{n-2}x^{n-2} + \ldots a_1x + a_0$$

where $a_n \neq 0$, *n* is a positive integer, and a_n, a_{n-1}, a_{n-2}, $\ldots a_1$, a_0 are constants. The coefficient of the greatest power of x, a_n, is called the **leading coefficient**. In the polynomial function $f(x) = 2x^4 - x^2 + 7x - 5$, the leading coefficient is 2. Since the term x^3 is missing, its coefficient is 0. Thus, $a_4 = 2$, $a_3 = 0$, $a_2 = -1$, $a_1 = 7$, and $a_0 = -5$.

Synthetic division: To divide $P(x) = 4x^3 - 7x^2 - 16x + 5$ by $x - 3$ using synthetic division, write the synthetic divisor 3 followed by the coefficients of the powers of x using the format illustrated below:

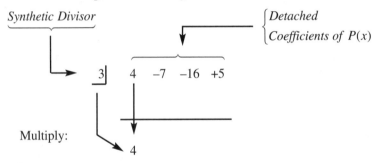

Bring down the first detached coefficient, 4, as shown above, and multiply it by the synthetic divisor, 3. Write the product, +12, above the horizontal line in the next column. Repeat this process as illustrated in the table that follows:

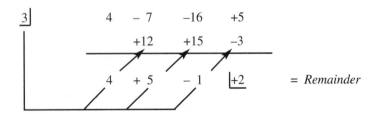

Reading the last line of the synthetic division table from left to right gives the detached coefficients of the second-degree polynomial quotient in standard form and the constant remainder. The quotient is $4x^2 + 5x - 1$ with remainder 2. Thus,

$$\frac{4x^3 - 7x^2 - 16x + 5}{x - 3} = 4x^2 + 5x - 1 + \frac{2}{x - 3}.$$

Guidelines for synthetic division: In dividing the polynomial $P(x)$ synthetically,

- Arrange the terms of $P(x)$ in standard form. If a power of x is missing from $P(x)$, then a coefficient of 0 is used in that position.
- If the divisor has the form $x - r$, use r as the synthetic divisor; if the divisor has the form $x + r$, the synthetic divisor is $-r$.
- The quotient obtained will always be a polynomial of degree one less than the degree of $P(x)$.

KEY EXAMPLE

Divide $x^4 + 2x^3 - 5x + 1$ by $x + 3$.

Solution: Provide for the missing power of x in the dividend by writing $x^4 + 2x^3 + 0\,x^2 - 5x + 1$. Put the divisor in the form $x - r$ by writing $x - (-3)$ so the synthetic divisor is -3.

$$
\begin{array}{r|rrrrr}
-3 & 1 & +2 & +0 & -5 & +1 \\
   &   & -3 & +3 & -9 & +42 \\
\hline
   & 1 & -1 & +3 & -14 & +43 \quad = \text{Remainder} \\
\end{array}
$$

$$\text{Quotient} = \quad x^3 \quad -x^2 \quad +3x \quad -14$$

Thus,

$$\frac{x^4 + 2x^3 - 5x + 1}{x + 3} = x^3 - x^2 + 3x - 14 + \frac{43}{x + 3}$$

KEY EXAMPLE

Divide $4x^3 - 8x^2 + x + 11$ by $2x - 3$.

Solution: Since the divisor can be written as $2(x - 3/2)$, divide $4x^3 - 8x^2 + x + 11$ synthetically by $x - 3/2$ and then divide the quotient by 2:

$$
\begin{array}{r|rrrr}
\frac{3}{2} & 4 & -8 & +1 & +11 \\
 & & +6 & -3 & -3 \\
\hline
 & 4 & -2 & -2 & \boxed{+8} \quad = \ \textit{Remainder}
\end{array}
$$

Dividing the detached coefficients of the quotient by 2 gives a quotient of $2x^2 - x - 1$ with a remainder of 8. Hence,

$$\frac{4x^3 - 8x^2 + x + 11}{2x - 3} = 2x^2 - x - 1 + \frac{8}{2x - 3}$$

KEY EXAMPLE

If $x - 2$ is a factor of $P(x) = x^3 + 2x^2 - 5x - 6$, factor $P(x)$ completely.

Solution: Divide $P(x)$ by $x - 2$ synthetically to obtain the corresponding quadratic factor of $P(x)$.

$$
\begin{array}{r|rrrr}
2 & 1 & +2 & -5 & -6 \\
 & & +2 & +8 & +6 \\
\hline
 & 1 & +4 & +3 & \boxed{+0}
\end{array}
$$

Since the quotient is $x^2 + 4x + 3$,

$$x^3 + 2x^2 - 5x - 6 = (x - 2)(x^2 + 4x + 3)$$

Factor $(x^2 + 4x + 3)$: $\qquad = (x - 2)(x + 1)(x + 3).$

Key 41 The remainder and factor theorems

OVERVIEW *The remainder obtained in dividing* P(x) *by* x − r *synthetically can be obtained also by evaluating* P(r). *If* P(r) = 0, *then* P(x) *is evenly divisible by* x − r *so* x − r *is a factor of* P(x).

The remainder theorem: If a polynomial $P(x)$ is divided by $x − r$, then the remainder is equal to the function value $P(r)$. In Key 40 we divided $P(x) = 4x^3 − 7x^2 − 16x + 5$ by $x − 3$ and obtained 2 as the remainder. This remainder, but not the quotient, may be obtained also by evaluating $P(3)$. Since $P(3) = 4(3^3) − 7(3^2) − 16(3) + 5 = 2$, the remainder is 2.

The factor theorem: A polynomial $P(x)$ has $x − r$ as a factor if and only if $P(r) = 0$. For example, to show that $x + 1$ is a factor of $P(x) = 2x^3 − 9x^2 + x + 12$, write $x + 1$ as $x − (−1)$ and evaluate $P(−1)$. Since $P(−1) = 2(−1)^3 − 9(−1)^2 + (−1) + 12 = 0$, $x + 1$ is a factor of $P(x)$. To find the polynomial that is the corresponding factor of $x + 1$, we would need to divide $P(x)$ by $x + 1$, as was done in the example in Key 40 (page 114).

KEY EXAMPLE

For what value of k is $(x + 3)$ a factor of $x^4 + x^3 − x^2 + kx − 12$?

Solution: The binomial $x + 3$ can be written as $x − (−3)$. Thus, $x − (−3)$ is a factor of $P(x) = x^4 + x^3 − x^2 + kx − 12$ if $P(−3) = 0$. Replacing x with $−3$ in $P(x)$ and setting the result equal to 0 gives

$$(−3)^4 + (−3)^3 − (−3)^2 + k(−3) − 12 = 0$$
$$81 − 27 − 9 − 3k − 12 = 0$$
$$33 − 3k = 0$$
$$k = \frac{−33}{−3} = 11$$

Thus, $(x + 3)$ is a factor of $x^4 + x^3 − x^2 + 11x − 12$ when $k = 11$.

Key 42 Zeros of polynomial functions

OVERVIEW *Any value of* x *for which a polynomial* P(x) *equals* 0 *is a **zero** of the function and a **root** of the polynomial equation,* P(x) = 0.

Zero of a polynomial function $P(x)$: A zero of a polynomial function is any number r, real or complex, for which $P(r) = 0$.

Fundamental theorem of algebra: Every nonconstant polynomial function has at least one complex zero.

Polynomial factorization theorem: A nonconstant polynomial $P(x)$ of degree n can be factored into n linear binomial factors. Thus, there exist numbers $r_1, r_2, r_3, \ldots, r_n$, real or complex, such that

$$P(x) = c\,(x - r_1)\,(x - r_2)\,(x - r_3) \ldots (x - r_n)$$

where c is a nonzero constant. The numbers $r_1, r_2, r_3, \ldots, r_n$ are the zeros of $P(x)$.

Multiplicity of zeros: The polynomial factorization theorem does not guarantee that the numbers $r_1, r_2, r_3, \ldots, r_n$ are all different. If the binomial $(x - r)$ appears as a factor k times, then r is said to be a **zero (root) of multiplicity k**. For example, if

$$P(x) = (x - 3)^2\,(x + 1)^3\,(x - 5),$$

then 3 is a zero (or root) of *multiplicity 2* since $x - 3$ appears as a factor *two* times. Thus, -1 is a zero (or root) of *multiplicity 3*. The polynomial $P(x)$ has 5 as a **simple zero** since its multiplicity is 1.

Making mathematical connections: The function $f(x) = x^3 - 3x^2 - 6x + 8$ can be looked at in three different ways.

- **Consider the function numerically**. The table-building feature of a graphing calculator can be used to quickly display the function values of f, as shown in Figure 7.1. Because $y = 0$ when $x = -2$, $x = 1$, and $x = 4$, these three values of x are the **zeros** of function f and the **roots** of the equation $x^3 - 3x^2 - 6x + 8 = 0$.

Figure 7.1 Table for $y = x^3 - 3x^2 - 6x + 8$

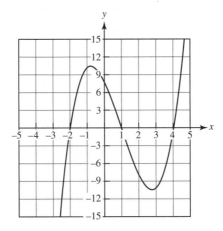

Figure 7.2 Graph of $f(x) = x^3 - 3x^2 - 6x + 8$

- **Consider the function graphically**. Because the graph of $y = f(x) = x^3 - 3x^2 - 6x + 8$ has intercepts at $(-2, 0)$, $(1, 0)$, and $(4, 0)$, as shown in Figure 7.2, -2, 1, and 4 are the **zeros** of function f and the **roots** of the equation $x^3 - 3x^2 - 6x + 8 = 0$.
- **Consider the function algebraically**. Function f can be written in factored form as

$$\begin{aligned} f(x) &= x^3 - 3x^2 - 6x + 8 \\ &= (x - 1)(x - 4)(x + 2) \\ &= (x - 1)(x - 4)(x - (-2)) \end{aligned}$$

Each linear factor of $f(x)$ is the difference between x and a zero of function f. Setting each linear factor equal to 0 gives a root of the equation $x^3 - 3x^2 - 6x + 8 = 0$.

Making equivalent statements: If any one of the following statements is true, then so is each of the other statements:

- k is a zero of function f.
- $x = k$ is a root of the equation $f(x) = 0$.
- $x - k$ is a factor of $f(x)$.
- $(k, 0)$ is an x-intercept of the graph of $y = f(x)$, provided that k is a real number.

n-Roots theorem of polynomial equations: A polynomial equation of degree n has exactly n roots, real or complex, provided that a root of multiplicity k is counted k times.

Although its roots may be difficult to find, the *fourth*-degree polynomial equation $x^4 + x^3 - 5x^2 + 2x + 7 = 0$ has a total of *four* roots. The fifth-degree equation $(x^2 + 1)(x - 6)^3 = 0$ has a total of *five* roots. Two of these roots are the imaginary numbers $\pm i$ which correspond to the roots of the equation $x^2 + 1 = 0$. Although the equation has only one distinct real root, namely, 6, it is a root of multiplicity 3 so it is counted three times.

Roots that occur in pairs: Nonreal solutions of polynomial equations with real coefficients occur in conjugate pairs. In other words, if $a + bi$ is a root of a polynomial equation (where $i = \sqrt{-1}$), then $a - bi$ is also a root. If a polynomial equation with rational coefficients has irrational roots, then they also occur in conjugate pairs. In other words, if $a + \sqrt{b}$ is a root of the equation, then $a - \sqrt{b}$ is also a root.

KEY EXAMPLE

Find the polynomial equation $P(x) = 0$ of lowest degree and with rational coefficients if its solution set includes $2i$, $- \sqrt{5}$, and 3 as a root of multiplicity 2.

Solution: For each root r of $P(x) = 0$, determine the corresponding factor $x - r$. If $2i$ is a root of $P(x) = 0$, then its conjugate, $-2i$, is also a root. Hence, $(x - 2i)$ and $(x + 2i)$ are factors of $P(x)$. Similarly, $(x + \sqrt{5})$ and $(x - \sqrt{5})$ are also factors of $P(x)$. Since 3 is a root of multiplicity 2, $(x - 3)^2$ is also a factor of $P(x)$. Thus, the equation in factored form is

$$(x - 2i)(x + 2i)(x + \sqrt{5})(x - \sqrt{5})(x - 3)^2 = 0.$$

Multiplying the factors together gives the sixth-degree equation

$$x^6 - 6x^5 + 8x^4 + 6x^3 - 29x^2 + 120x - 180 = 0.$$

Key 43 Reducing the degree of a polynomial equation

OVERVIEW *The degree of a polynomial equation is reduced by one each time the equation is divided by a factor that corresponds to a root. By removing a sufficient number of roots from a polynomial equation, it may be possible to obtain a quadratic equation.*

Depressed equation: If $x - r$ is a factor of an nth-degree polynomial $P(x)$, then the equation obtained by dividing $P(x)$ by $x - r$ and setting the quotient equal to 0 is an equation of degree $n - 1$, called the ***depressed equation***. If we know one or more roots of a polynomial equation, it may be possible to obtain a second-degree depressed equation, which can be solved either by factoring or, if necessary, by using the quadratic formula.

KEY EXAMPLE

Given the equation $P(x) = x^3 - 13x + 12 = 0$, find one root by inspection and the other roots by solving the depressed equation.

Solution: By inspecting the equation $P(x) = x^3 - 13x + 12 = 0$, we see that $P(1) = 0$, so $x = 1$ is a solution. The depressed equation is obtained by dividing $x^3 - 13x + 12$ by $x - 1$ synthetically and setting the quotient equal to 0. This gives $x^2 + x - 12 = 0$ whose roots are 3 and –4.

Thus, the three roots of the original equation are 1, 3, and –4.

KEY EXAMPLE

If two roots of the equation $P(x) = x^4 + x^3 - 15x^2 - 22x + 8 = 0$ are –2 and 4, find the remaining roots.

Solution: Divide $P(x)$ by $x - (-2)$ and then divide the quotient obtained by $x - 4$.

$$\begin{array}{r|rrrrr} -2 & 1 & 1 & -15 & -22 & +8 \\ & & -2 & +2 & -26 & -8 \\ \hline \end{array}$$

$$\begin{array}{r|rrrr|r} 4 & 1 & -1 & -13 & +4 & \underline{0} \\ & & +4 & +12 & -4 & \\ \hline \end{array}$$

$$\begin{array}{rrrr|r} & 1 & +3 & -1 & \underline{0} \end{array}$$

Quotient: $x^2 \quad +3x \quad -1$

Depressed Equation: $x^2 \quad +3x \quad -1 \quad = 0$

The roots of the depressed equation can be obtained using the quadratic formula:

$$x = \frac{-3 \pm \sqrt{9 - 4(1)(-1)}}{2} = \frac{-3 \pm \sqrt{13}}{2}.$$

Hence, the four roots of the original equation are -2, 4, and $\dfrac{-3 \pm \sqrt{13}}{2}$.

Key 44 The rational root theorem

OVERVIEW *The leading coefficient and the constant term of a polynomial equation provide clues to any possible rational roots of the equation.*

Integer root theorem: If an integer p is a root of a polynomial equation having integer coefficients and a leading coefficient of 1, then p must be a factor of the constant term of the equation. Since the equation $x^4 + x^3 - 15x^2 - 22x + 8 = 0$ has integer coefficients and its leading coefficient is 1, any integer roots must be factors of 8. Thus, the possible integer roots of this equation are limited to ± 1, ± 2, ± 4, and ± 8.

Rational root theorem: If p/q is a rational root in lowest terms of the polynomial equation with integer coefficients

$$a_n x^n + a_{n-1} x^{n-1} + a_{n-2} x^{n-2} + \cdots + a_1 x + a_0 = 0 \quad \text{(where } a_n \neq 0\text{)},$$

then p is a factor of a_0 and q is a factor of a_n.

Example. The equation $P(x) = 3x^4 - 7x^3 + 8x^2 - 14x + 4 = 0$ may or may not have rational roots. If the equation does have rational roots, then each rational root must be a fraction whose numerator is a factor of 4 and whose denominator is a factor of 3. Since the factors of 4 are ± 1, ± 2, and ± 4, and the factors of 3 are ± 1 and ± 3, the possible rational roots of the equation are limited to

$$\pm \frac{1}{3}, \ \pm \frac{2}{3}, \ \pm 1, \ \pm \frac{4}{3}, \ \pm 2, \text{ and } \pm 4 \ .$$

Using synthetic division or the remainder theorem, we can systematically test each possible rational root to see which of these, if any, are actually roots of the given equation. Since $P(2) = 0$ and $P(1/3) = 0$, 2 and 1/3 are rational roots of the equation. Successively dividing $P(x)$ by the binomial factors that correspond to each root gives a second-degree depressed equation:

1/3	3	− 7	8	−14	+4
	↓	+ 1	−2	+ 2	−4

2	3	− 6	+6	−12	0
	↓	+6	+02	+12	

	3	0	+6	0

The last depressed equation is $3x^2 + 6 = 0$ whose roots are $\pm i\sqrt{2}$. Hence, the four roots of the original equation are 1/3, 2, and $\pm i\sqrt{2}$.

Key 45 Locating real zeros

OVERVIEW   *If a polynomial function* P(x) *changes sign in the interval* (a, b), *then a **real zero** of* P(x) *is located between* x = a *and* x = b.

Finding real zeros of polynomial functions: Locate $(1, -2)$ and $(2, 3)$ on the graph of $f(x) = -x^3 + 4x^2 - 5$ given in Figure 7.3. As x varies from 1 to 2, function f will assume every real number from $f(1) = -2$ to $f(2) = 3$. The generalization of this observation is called the **intermediate value theorem**. Because $f(1)$ and $f(2)$ have opposite signs, the graph crosses the x-axis at some point between $x = 1$ and $x = 2$. Thus, whenever $f(a)$ and $f(b)$ have opposite signs, $f(x)$ has a real zero between $x = a$ and $x = b$.

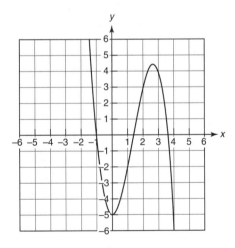

Figure 7.3 Graph of $f(x) = -x^3 + 4x^2 - 5$

KEY EXAMPLE

In the interval $[-2, 4]$, find any pairs of consecutive integers between which the zeros of $p(x) = 6x^4 - 13x^3 + 7x^2 - 26x - 10$ are located.

Solution: Use the table-building feature of your graphing calculator with $Y1 = 6x^4 - 13x^3 + 7x^2 - 26x - 10$, as shown in the accompanying figure.

X	Y1	
-2	270	
-1	42	
0	-10	
1	-36	
2	-42	
3	110	
4	702	

X=4

- The sign of *Y*1 changes from positive to negative as *x* increases from *x* = –1 to *x* = 0, so there is at least one real zero between –1 and 0.
- The sign of *Y*1 changes from negative to positive as *x* increases from *x* = 2 to *x* = 3, so there is at least one real zero between 2 and 3.

Determining the number of real zeros: If $f(x)$ changes sign between two different *x*-values, say *x* = 1 and *x* = 2, does that change indicate that there is exactly one zero between 1 and 2? If $f(x)$ does *not* change sign between two different *x*-values, does the lack of change mean that there are no zeros between those *x*-values? In Figure 7.4, $f(0)$ and $f(1)$ have the *same* sign, and *f* has two zeros between 0 and 1; $f(1)$ and $f(2)$ have *opposite* signs, and there are three zeros between 1 and 2; and $f(3)$ have the *same* sign, and *f* has no zeros between 2 and 3.

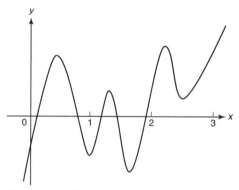

Figure 7.4 The Intermediate Value Theorem

Intermediate value theorem: Let *f* represent a polynomial function such that $f(a) \neq f(b)$, where *a* and *b* stand for real numbers with *a* < *b*.

- In the interval from *x* = *a* to *x* = *b*, *f* takes on every number between $f(a)$ and $f(b)$.
- If $f(a)$ and $f(b)$ have *opposite* signs, there is an odd number of real zeros of *f* between *a* and *b*.
- If $f(a)$ and $f(b)$ have the *same* sign, there are no real zeros of *f* between *a* and *b* *or* there is an even number of real zeros.

Key 46 Rational functions and asymptotes

OVERVIEW *A polynomial function in* x *contains only positive-integer powers of* x. *A rational function in* x *is the quotient of two polynomial functions in* x.

Definition of rational function: If $P(x)$ and $Q(x)$ are polynomial functions, then $f(x) = \dfrac{P(x)}{Q(x)}$ is a rational function the domain of which excludes those values of x, if any, for which $Q(x) = 0$.

Point discontinuities:

If $f(x) = \dfrac{P(x)}{Q(x)}$ and $P(x)$ and $Q(x)$ have $x - c$ as a common factor, then the graph of $f(x)$ has a point discontinuity or "hole" at $x = c$. For example, since

$$f(x) = \frac{x^2 - 4}{x - 2}$$

$$= \frac{(x - 2)(x + 2)}{x - 2}$$

$$= x + 2$$

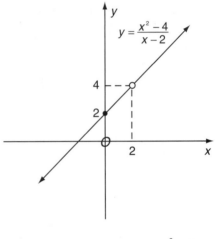

Figure 7.5 Graph of $f(x) = \dfrac{x^2 - 4}{x - 2}$

the graph of $f(x)$ has a point discontinuity at $x = 2$. Thus, the graph of $y = f(x)$ is identical to the graph of $y = x + 2$ except that an open circle at $x = 2$ indicates that the graph of $y = \dfrac{x^2 - 4}{x - 2}$ does not contain point (2, 4). See Figure 7.5.

Asymptote: An asymptote is a line that a curve approaches, but does not intersect, as the curve is extended indefinitely. A curve may have vertical asymptotes, horizontal asymptotes, or both types of asymptotes.

- A line $x = a$ is a **vertical asymptote** of the graph of $y = f(x)$ if, as $x \to a$ (read as "x approaches a"), $f(x) \to \infty$ or $f(x) \to -\infty$ (read as "f of x approaches infinity" or "f of x approaches negative infinity"). The expression "approaches infinity" means increases without having an upper bound, while the expression "approaches negative infinity" means decreases without having a lower bound. In Figure 7.6 the y-axis (the line $x = 0$) is a vertical asymptote of the graph of $f(x) = \dfrac{1}{x^2}$ since, as x approaches 0 from either the positive or the negative side of 0, the graph of $f(x)$ constantly rises while getting closer and closer to the y-axis without ever touching it.

- A line $y = b$ is a **horizontal asymptote** of the graph of $y = f(x)$ if, as $x \to \infty$ or as $x \to -\infty$, $f(x) \to b$. In Figure 7.6, the graph of $f(x) = \dfrac{1}{x^2}$, the x-axis (the line $y = 0$) has a horizontal asymptote since, as x increases (or decreases) without bound, the graph of $f(x)$ constantly falls while getting closer and closer to the x-axis without ever touching it.

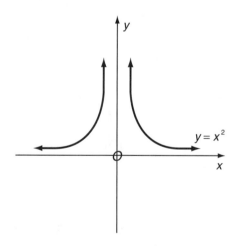

Figure 7.6 Graph of $f(x) = \dfrac{1}{x^2}$

Key 47 Graphing rational functions

OVERVIEW *A rational function* f *has the form*
$f(x) = \dfrac{P(x)}{Q(x)}$*, where* $P(x)$ *and* $Q(x)$ *are polynomial functions.*
The graph of a rational function may have asymptotes as x
approaches the values for which the function is undefined.
When discussing the behavior of a rational function, it is
helpful to know any intercepts and asymptotes.

Finding intercepts of $f(x) = P(x)/Q(x)$**:** To determine the points at
which the graph of $f(x) = \dfrac{P(x)}{Q(x)}$ intersects the coordinate axes:

- Set $P(x) = 0$. Any solution of this equation that does not also make
$Q(x)$ equal to 0 is an x-intercept.
- Set $x = 0$ and evaluate $f(0)$. The y-intercept is $(0, f(0))$, provided
that this value is defined.

Example 1: If $f(x) = \dfrac{x^2 - 16}{x - 8}$, then the x-intercepts are the roots of

the equation $x^2 - 16 = 0$ provided that these roots, $x = \pm 4$, do not make
the denominator evaluate to 0. Since the denominator is not 0 when
$x = \pm 4$, the x-intercepts are $(-4, 0)$ and $(4, 0)$.

Example 2: If $f(x) = \dfrac{x^2 - 16}{x - 8}$, then

$$f(0) = \frac{0^2 - 16}{0 - 8} = -2, \text{ so the } y\text{-intercept is } (0, -2).$$

Finding vertical asymptotes of $f(x) = \dfrac{P(x)}{Q(x)}$: The graph of a rational

function has a *vertical* asymptote at each value of x, if any, for
which the denominator evaluates to 0 but the numerator does not
evaluate to 0. The line $x = a$ is a vertical asymptote of the graph of
$f(x) = \dfrac{P(x)}{Q(x)}$ when $Q(a) = 0$ and $P(a) \neq 0$.

Example: To find any vertical asymptotes of $f(x) = \dfrac{4x + 8}{x^2 - x - 6}$,

rewrite $f(x)$ in factored form as $f(x) = \dfrac{4(x + 2)}{(x - 3)(x + 2)}$. Although $x = 3$

and $x = -2$ both make the denominator of $f(x)$ equal 0, only $x = 3$ is a vertical asymptote. Because $x = -2$ makes the numerator and the denominator 0 at the same time, the graph will have a discontinuity at this point.

Finding a horizontal asymptote of $f(x) = \dfrac{P(x)}{Q(x)}$: The graph of $f(x) = \dfrac{P(x)}{Q(x)}$ has a *horizontal* asymptote when the degree of the numerator is less than or equal to the degree of the denominator. Let

$$f(x) = \frac{a_n x^n + a_{n-1}x^{n-1} + \cdots + a_1 x + b_0}{b_m x^m + b_{m-1}x^{m-1} + \cdots + b_1 x + b_0},$$

where $a_n \neq 0$ and $b_m \neq 0$.

- If $n < m$, the x-axis is a horizontal asymptote. The graph of $f(x) = \dfrac{3x}{x^2 - 9}$ has the x-axis as a horizontal asymptote since the degree of the numerator is less than the degree of the denominator. When checking for vertical asymptotes, we find that the graph has vertical asymptotes at $x = -3$ and $x = 3$ since, for these x-values, the denominator of the function is not defined. See Figure 7.7.

- If $n = m$, $y = \dfrac{a_n}{b_m}$ is a horizontal asymptote. Since the degrees of the numerator and the denominator of $f(x) = \dfrac{\boxed{3}x^2 - x - 4}{\boxed{2}x^2 - 5x + 6}$ are equal, $y = \dfrac{3}{2}$ is a horizontal asymptote of the graph, as shown in Figure 7.8.

- If $n > m$, f has no horizontal asymptote. The graph of $f(x) = \dfrac{x^3 - 2x + 5}{2x^2 + x - 4}$ does not have a horizontal asymptote.

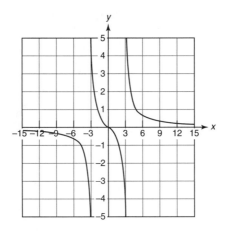

Figure 7.7 Graph of $f(x) = \dfrac{3x}{x^2 - 9}$

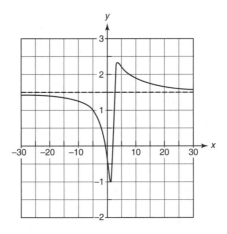

Figure 7.8 Graph of $f(x) = \dfrac{3x^2 - x - 4}{2x^2 - 5x + 6}$

Finding a slant asymptote of $f(x) = \dfrac{P(x)}{Q(x)}$: If the degree of the numerator of a rational function is one more than the degree of the denominator, the graph of the function has a **slant** asymptote. An equation of a slant asymptote of a rational function f can be obtained by dividing the numerator of function f by its denominator, deleting any remainder, and then setting y equal to the remaining part of the quotient. For example, to find an equation of the slant asymptote of the graph of

$$f(x) = \frac{3x^2 - 12x + 2}{x - 4} :$$

- Divide the numerator by the denominator:

$$
\begin{array}{r}
3x \\
x - 4 \overline{)3x^2 - 12x + 2} \\
-\ \underline{3x^2 - 12x} \\
+2
\end{array}
$$

The answer is $3x + \dfrac{2}{x - 4}$.

- Delete the remainder from the answer: $3x + \cancel{\dfrac{2}{x - 4}}$.

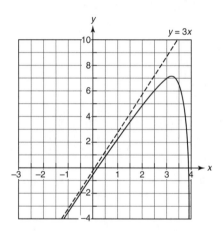

Figure 7.9 Graph of $f(x) = \dfrac{3x^2 - 12x + 2}{x - 4}$

- Write the equation of the slant asymptote by setting y equal to the remaining part of the quotient: $y = 3x$.

Figure 7.9 shows the graph of the function with its asymptote.

KEY EXAMPLE

Find the intercepts and asymptotes of the graph of $f(x) = \dfrac{x^2 - 36}{x^2 - 9}$. Discuss any symmetry that the graph displays.

Solution: Factor $f(x)$ as $f(x) = \dfrac{(x - 6)(x + 6)}{(x - 3)(x + 3)}$.

- **x-intercepts:** $(-6, 0)$ and $(6, 0)$ because $f(x) = 0$ when $(x = 6)$ or $x = -6$.

- **y-intercepts:** $(0, 4)$ because, if $x = 0$, $y = f(0) = \dfrac{0^2 - 36}{0^2 - 9} = 4$.

- **asymptotes**: $x = +3$ and $x = -3$ because these values of x make the denominator 0 but do not make the numerator 0.

- **symmetry:** $f(x)$ is an even function since $f(-x) = f(x)$. Then its graph is symmetric with respect to the y-axis, as shown in the accompanying figure, which was obtained using a graphing calculator set to a decimal window.

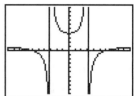

KEY EXAMPLE

Find an equation of the slant asymptote of the graph of $f(x) = \dfrac{x^2 - 16}{x + 8}$.

Solution: Using long division, divide the numerator by the denominator:

$$
\begin{array}{r}
x - 8 \\
x + 8 \overline{)\, x^2 - 16} \\
-\ \underline{x^2 + 8x} \\
-8x - 16 \\
\underline{-8x - 64} \\
48
\end{array}
$$

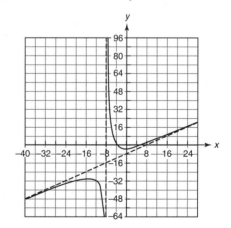

The answer is $x - 8 + \dfrac{48}{x + 8}$.

Deleting the remainder leaves $x - 8$. The equation of the slant asymptote is $y = x - 8$, as shown in the accompanying figure.

Key 48 Decomposing rational expressions

OVERVIEW *You already know how to find the sum of two fractions:*

$$\frac{2}{x} + \frac{5}{x-3} = \frac{7x-6}{x^2-3x} .$$

*Sometimes, as in the study of calculus, it is necessary to reverse this process by starting with the fraction on the right side of the equation and "decomposing it" by writing it as the sum or difference of two or more simpler fractions, called **partial fractions**.*

Irreducible quadratic factors: The quadratic expression $ax^2 + bx + c$ is *irreducible* if it cannot be factored over the set of real numbers into the product of linear factors. The quadratic $x^2 + 4$ is irreducible since its factors are not real, as in $(x - 2i)(x + 2i)$.

Linear-quadratic factorization theorem for polynomials: Every polynomial function with real coefficients can be uniquely factored over the real numbers into a product of linear or irreducible quadratic factors.

Proper versus improper rational expressions: The rational expression $\frac{7x-6}{x^2-3x}$ is proper. A rational expression is *proper* if the degree of the polynomial in the numerator is lower than the degree of the polynomial in the denominator. Otherwise, the rational expression is **improper**. Only proper fractions can be decomposed into the sum of two or more partial fractions.

The decomposition principle: The method used for *decomposing* a proper rational expression into partial fractions depends largely on the nature of the factors in the denominator. Because of the linear-quadratic factorization theorem, there are four cases to consider:

- The denominator contains only linear factors, all of which are different.
- The denominator contains at least one linear factor that is repeated.
- The denominator includes an irreducible quadratic factor.
- The denominator contains at least one irreducible quadratic factor that is repeated.

When studying the four cases, we assume that the rational expression is proper and is in lowest terms.

Case I (nonrepeated linear factor): If the denominator of a rational expression can be written as the product of different nonrepeated linear factors of the form $ax + b$, then, for each of these factors the decomposition of the rational expression includes a partial fraction

of the form $\dfrac{A}{ax+b}$, where A is some real number.

Example. To decompose $\dfrac{7x-6}{x^2-3x}$, factor the denominator and then

provide a partial fraction for each linear factor:

$$\frac{7x-6}{x(x-3)} = \frac{A}{x} + \frac{B}{x-3}.$$

Next, clear this equation of its fractions by multiplying each term by $x(x-3)$:

$$7x - 6 = A(x-3) + Bx$$
$$= Ax - 3A + Bx$$
$$= (A+B)x - 3A$$

Compare the coefficients of the x-terms on both sides of the equation, and compare the constant terms on both sides of the equation. It must be the case that $3A = 6$, so $A = 2$, and also that $A + B = 7$, so $B = 7 - A = 7 - 2 = 5$. Hence:

$$\frac{7x-6}{x^2-3x} = \frac{2}{x} + \frac{5}{x-3}.$$

Case II (repeated linear factor): If the factored form of the denominator of a rational expression includes a factor of the form $(ax + b)^m$, where m is a positive integer greater than 1, then the decomposition of the rational expression includes the sum of m partial fractions of the form

$$\frac{A_1}{ax+b} + \frac{A_2}{(ax+b)^2} + \cdots + \frac{A_m}{(ax+b)^m},$$

where $A_1, A_2, \ldots, A_m$ are m real-valued constants.

Example: To decompose $\dfrac{8x+7}{(2x-1)^3}$, provide three partial fractions

because the exponent of $2x - 1$ is 3:

$$\frac{8x+7}{(2x-1)^3} = \frac{A}{2x-1} + \frac{B}{(2x-1)^2} + \frac{C}{(2x-1)^3}.$$

Applying the decomposition principle: To find the partial fraction decomposition of $\dfrac{3x^2 - 29x + 80}{x(x-4)^2}$, follow these steps:

Step 1. Determine the partial fraction that corresponds to each factor of the denominator. The partial fraction $\dfrac{A}{x}$ corresponds to the factor x. For the factor $(x-4)^2$, there corresponds the partial fraction sum

$$\frac{B}{x-4} + \frac{C}{(x-4)^2} .$$

Step 2. Rewrite the original fraction in terms of its partial fraction decomposition:

$$\frac{3x^2 - 29x + 80}{x(x-4)^2} = \frac{A}{x} + \frac{B}{x-4} + \frac{C}{(x-4)^2} .$$

Step 3. Determine the values of the constants.

- Clear the equation of its fractional terms by multiplying each term by $x(x-4)^2$, the lowest common multiple of the denominators:

$$3x^2 - 29x + 80 = A(x-4)^2 + Bx(x-4) + Cx.$$

- On the right side of the equation, perform the indicated operations and then collect like terms:

$$\begin{aligned} 3x^2 - 29x + 80 &= A(x^2 - 8x + 16 + Bx(x^2 - 4x) + Cx \\ &= (A+B)x^2 + (-8A - 4B + C)x + 16A \end{aligned}$$

- Set the numerical coefficients of like variable terms on the two sides of the equation equal. Do the same for the constant terms:

x^2-coefficients:	$3 = A + B$
x-coefficients:	$-29 = -8A - 4B + C$
constant terms:	$80 = 16A$

- Solve the system of three equations in three unknowns by starting with the simplest of the three equations. Since $80 = 16A$,

$A = \dfrac{80}{16} = 5$.

Substitute $A = 5$ in the first equation; then $B = 3 - A = 3 - 5 = -2$. Substitute $A = 5$ and $B = -2$ in the second equation; then $C = 3$.

Step 4. Use the values of the constants to write the complete partial fraction decomposition:

$$\frac{3x^2 - 29x + 80}{x(x-4)^2} = \frac{5}{x} + \frac{-2}{x-4} + \frac{3}{(x-4)^2} .$$

Case III (nonrepeated irreducible quadratic factor): For each non-repeated irreducible quadratic factor of the form $ax^2 + bx + c$ that the denominator of a rational expression contains, the decomposition of that fraction requires a partial fraction of the form $\dfrac{Ax + B}{ax^2 + bx + c}$.

Example:

$$\frac{3x^2 + 7}{(x-1)(x^2 + 2x + 5)} = \frac{A}{x-1} + \frac{Bx + C}{x^2 + 2x + 5}$$

Case IV (repeated irreducible quadratic factor): For each repeated irreducible quadratic factor of the form $(ax^2 + bx + c)^m$ that the denominator of a proper fraction contains, where m is a positive integer greater than 1, the decomposition of that fraction requires a partial fraction of the form

$$\frac{A_1 x + B_1}{ax^2 + bx + c} + \frac{A_2 x + B_2}{\left(ax^2 + bx + c\right)^2} + \cdots + \frac{A_m x + B_m}{\left(ax^2 + bx + c\right)^m},$$

where $A_1, A_2, \ldots, A_m$ and $B_1, B_2, \ldots, B_m$ are real-valued constants.

Example:

$$\frac{x}{\left(x^2 + 1\right)^3} = \frac{A_1 x + B_1}{x^2 + 1} + \frac{A_2 x + B_2}{\left(x^2 + 1\right)^2} + \frac{A_3 x + B_3}{\left(x^2 + 1\right)^3}$$

Improper rational expressions: If the original rational expression is improper, use long division to rewrite it as the sum of a polynomial (the quotient) and a proper fraction (the remainder). Then decompose the proper fraction. Before you can apply the decomposition principle to $\dfrac{2x^2 + x - 6}{x^2 - 3x}$, you must rewrite it as the sum of a polynomial and a proper fraction. Use long division to verify that

$$\frac{2x^2 + x - 6}{x^2 - 3x} = 2 + \frac{7x - 6}{x^2 - 3x}.$$

From the example on page 133, you already know that

$$\frac{7x - 6}{x^2 - 3x} = \frac{2}{x} + \frac{5}{x - 3}.$$

Hence, $\dfrac{2x^2 + x - 6}{x^2 - 3x} = 2 + \dfrac{2}{x} + \dfrac{5}{x - 3}.$

Theme 8 EXPONENTIAL AND
LOGARITHMIC FUNCTIONS

*M*ost of our work with functions has been limited to algebraic functions. **Algebraic functions** are functions that can be defined in terms of the sums, differences, products, quotients, or roots of polynomial functions. Nonalgebraic functions often arise in attempts to represent real-life phenomena mathematically. **Exponential** and **logarithmic functions** are important types of nonalgebraic functions.

INDIVIDUAL KEYS IN THIS THEME

Key 49 Exponential functions

OVERVIEW *The function* f(x) = x³ *is an example of a polynomial function. If the roles of the variable base* x *and the constant power* 3 *are interchanged, then the function that results,* F(x) = 3ˣ, *is called an **exponential** function.*

Exponential function: An exponential function has the form $f(x) = b^x$ where *b* is a positive constant different from 1. The domain of *x* is the set of real numbers. Since $f(x)$ has a positive value for all possible replacements of *x*, the range of an exponential function is limited to the set of positive real numbers.

Graphs of exponential functions: In Figure 8.1, we can compare the graphs of $y = b^x$ when *b* > 1 and when 0 < *b* < 1. From the graphs, we observe the following:

- For *b* > 1, the function $y = b^x$ is an *increasing function,* rising as *x* increases. For 0 < *b* < 1, $y = b^x$ is a *decreasing function*, falling as *x* increases.
- For *b* > 1 and for 0 < *b* < 1, the graphs intersect the *y*-axis at (0, 1) but *do not* intersect the *x*-axis.
- The *x*-axis is an asymptote. For *b* > 1, the curve approaches the negative *x*-axis as *x* gets smaller. For 0 < *b* < 1, the curve approaches the positive *x*-axis as *x* gets larger.

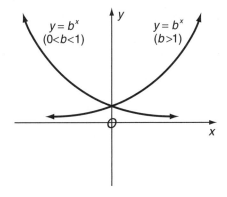

Figure 8.1 Graphs of $y = b^x$

Graphs of $y = b^x$ and $y = b^{-x}$ ($b > 0$ and $b \neq 1$): The graphs of $y = b^x$ and $y = b^{-x} = \left(\dfrac{1}{b}\right)^x$ are reflections of each other in the y-axis.

Figure 8.2 shows the graphs of $y = 2^x$ and $y = 2^{-x} = \left(\dfrac{1}{2}\right)^x$.

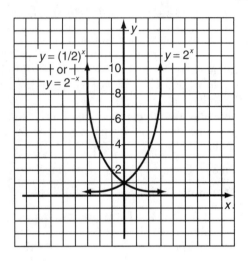

Figure 8.2 Graphs of $y = 2^x$ and $y = 2^{-x}$

One-to-one property of exponential functions: The graph of an exponential function $f(x) = b^x$ passes the horizontal line test so it is a one-to-one function. Thus, if $b^{x_1} = b^{x_2}$, then $x_1 = x_2$.

Exponential equation: An exponential equation is one in which the variable appears as part of an exponent. Some exponential equations can be solved by expressing each side as a power of the same base and then equating the exponents.

Example: If

$$8^{2x} = 16^{x+3},$$

then substituting 2^3 for 8 and 2^4 for 16 gives

$$2^{6x} = 2^{4x+12}.$$

By the one-to-one property of exponential functions, the exponents must be equal so

$$6x = 4x + 12$$

and

$$x = 6.$$

The natural base e: This special irrational number is denoted by the letter e in honor of the eighteenth-century mathematician Leonhard Euler. The number e is approximately equal to 2.718 and can be estimated to any desired level of accuracy using the expression $\left(1 + \dfrac{1}{n}\right)^n$ for which larger and larger values of n give increasingly accurate approximations of e.

Natural exponential function: An exponential function whose base is e is a natural exponential function. These functions arise when certain types of growth processes are represented mathematically.

Key 50 Logarithmic functions

OVERVIEW *The inverse of the exponential function* $y = b^x$ *is the function* $x = b^y$. *In order to solve for* y *in terms of* x, *the **logarithmic function** was invented. If* $x = b^y$, *then* $y = \log_b x$, *which is read as "**logarithm of x to base b is y**."*

Definition of logarithmic function: The **logarithm** of a *positive* number x to a given base b, written as $\log_b x$, is the power to which b must be raised to obtain x, provided that $b > 0$ and $b \neq 1$. A logarithm is an exponent since

$$y = \log_b x \text{ means } b^y = x.$$

Graphs of logarithmic functions: Figures 8.3 and 8.4 compare the graphs of $y = b^x$ and its inverse function, $y = \log_b x$, for different values of their base. Notice that either graph is the reflection of the other in the line $y = x$. The accompanying table compares the features of these graphs.

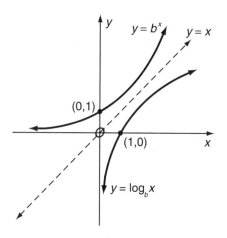

Figure 8.3 Graphs of $y = b^x$ and $y = \log_b x$ $(b > 1)$

Comparing the graphs of $y = b^x$ and $y = \log_b x$

Feature	Graph of $y = b^x$	Graph of $y = \log_b x$
Quadrants of graph	I and II	I and IV
Domain	all real numbers	positive real numbers
Range	positive real numbers	all real numbers
Intercepts	(0, 1)	(1, 0)
Asymptote	x-axis	y-axis
$b > 1$	rises	rises
$0 < b < 1$	falls	falls

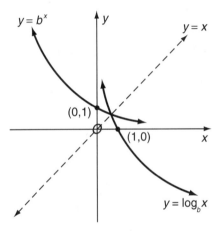

Figure 8.4 Graphs of $y = b^x$ and $y = \log_b x$ $(0 < b < 1)$

Equivalent exponential and logarithmic forms: To convert one form to the other, we use the fact that $y = \log_b x$ and $x = b^y$ mean the same thing. Here are some examples:

Exponential Form	Logarithmic Form
$4^3 = 64$	$\log_4 64 = 3$
$\sqrt{25} = 5$	$\log_{25} 5 = \dfrac{1}{2}$
$7^0 = 1$	$\log_7 1 = 0$
$3^{-2} = \dfrac{1}{9}$	$\log_3 \dfrac{1}{9} = -2$

Common logarithms: A common logarithm is a logarithm whose base is 10. A common logarithm is indicated when the base is omitted. Thus, $\log x$ means $\log_{10} x$. The common logarithm of a power of 10 is simply the exponent of 10. Thus, $\log 10 = 1$ since $10 = 10^1$; $\log 100 = 2$ since $100 = 10^2$; $\log 1000 = 3$ since $1000 = 10^3$; and so forth.

Natural logarithm: A natural logarithm, denoted by $\ln x$, is a logarithm whose base is e. Thus, $\ln x = \log_e x$. Here are some additional properties of the natural log function:

- $\ln e = 1$ since, in exponential form, $e^1 = e$.
- $\ln (e^x) = e^{\ln x} = x$ since the natural log and natural exponential functions are inverse functions.

KEY EXAMPLE

Solve for x: (a) $\log_x 8 = \dfrac{3}{2}$ (b) $\log_9(27x) = x + 1$

Solution: Change each equation into exponential form and then solve the resulting equation.

(a) If
$$\log_x 8 = \frac{3}{2},$$
then
$$x^{3/2} = 8$$
and
$$x = 8^{2/3} = 4.$$

(b) If
$$\log_9(27^x) = x + 1,$$
then
$$9^{x+1} = 27^x$$
$$3^{2(x+1)} = 3^{3x}$$
$$2(x+1) = 3x$$
$$x = 2$$

KEY EXAMPLE

Solve for x: $\ln e^{\sqrt{x}} = 2$

Solution: Since $\ln(e^x) = x$,
$$\ln e^{\sqrt{x}} = \sqrt{x} = 2.$$
Thus,
$$x = 4.$$

Key 51 Logarithm laws

OVERVIEW *Since logarithms are exponents, the laws for finding the logarithms of a product, quotient, and power are consistent with the laws of exponents for these operations.*

Product law of logarithms: The logarithm of a product equals the sum of the logarithms of its factors. Thus,

$$\log_b(xy) = \log_b x + \log_b y.$$

Example:

$$\log_4 21 = \log_4(7 \cdot 3) = \log_4 7 + \log_4 3$$

Quotient law of logarithms: The logarithm of a quotient equals the logarithm of the numerator minus the logarithm of the denominator. Thus,

$$\log_b\left(\frac{x}{y}\right) = \log_b x - \log_b y.$$

Example:

$$\log\left(\frac{pq}{100}\right) = \log(pq) - \log 100 = \log p + \log q - 2$$

Power law of logarithms: The logarithm of a power of a quantity equals the product of the exponent and the logarithm of the quantity. Thus,

$$\log_b x^n = n \log_b x.$$

Example 1:

$$\log \sqrt{10x} = \log(10x)^{1/2} = \frac{1}{2}\log 10x$$

$$= \frac{1}{2}(\log 10 + \log x)$$

$$= \frac{1}{2}(1 + \log x)$$

Example 2:

$$\ln e^2 + \ln \sqrt[3]{e} = 2\ln e + \frac{1}{3}\ln e$$

$$= 2(1) + \frac{1}{3}(1)$$

$$= \frac{7}{3}$$

KEY EXAMPLE

If $x = \log m$ and $y = \log n$, express $\log \sqrt[3]{\dfrac{m}{n}}$ in terms of x and y.

Solution:

$$\log \sqrt[3]{\frac{m}{n}} = \frac{1}{3}(\log m - \log n) = \frac{1}{3}(x - y)$$

KEY EXAMPLE

If $\log 2 = x$ and $\log 3 = y$, express each of the following in terms of x and y: (a) $\log 1.5$ (b) $\log \sqrt[3]{18}$

Solutions: (a) $\log 1.5 = \log \dfrac{3}{2} = \log 3 - \log 2 = y - x.$

(b) $\log \sqrt[3]{18} = \log(2 \cdot 3^2)^{\frac{1}{3}} = \dfrac{1}{3}(\log 2 + 2\log 3) = \dfrac{1}{3}(x + 2y)$

KEY EXAMPLE

If $\log N = 2 \log x + \log y$, express N in terms of x and y.

Solution: Work "backwards" in applying the logarithm laws for powers and products:

$$\log N = 2 \log x + \log y = \log x^2 + \log y = \log x^2 y.$$

Since the logarithm function is one-to-one, $\log A = \log B$ implies $A = B$. Hence, $\log N = \log x^2 y$ means $N = x^2 y$.

Formula for a change of logarithm base: To change $\log_a c$ to an equivalent expression whose logarithm base is b, use the change of base formula

$$\log_a c = \frac{\log_b c}{\log_b a}.$$

For example:

- To express $\ln 7$ to a base 10 logarithm, let $a = e$, $c = 7$, and $b = 10$. Then

$$\ln 7 = \frac{\log_{10} 7}{\log_{10} e} = \frac{\log 7}{\log e}.$$

- To use a graphing calculator to graph $y = \log_3 x$, set

$$Y_1 = \frac{\log(x)}{\log(3)}.$$

Key 52 Exponential and logarithmic

equations

OVERVIEW *Logarithm laws and properties can be used to solve exponential equations and equations that contain more than one logarithmic expression.*

Solving exponential equations using logarithms: If it is not possible to express both sides of an exponential equation as the power of the same rational base, as in $2^{3x} = 6$, then take the logarithm of each side of the equation:

$$\log\left(2^{3x}\right) = \log 6$$
$$3x \cdot \log 2 = \log 6$$
$$x = \frac{\log 6}{3\log 2} \approx 0.8618$$

KEY EXAMPLE

Solve $3^x = 21$. Approximate x to the *nearest hundredth*. Confirm your answer graphically.

Solution: If $3^x = 21$, then $\log(3^x) = \log 21$, so

$$x \log 3 = \log 21 \quad \text{and} \quad x = \frac{\log 21}{\log 3} \approx 2.77.$$

To confirm your answer graphically:
- Set $Y_1 = 3 \wedge x$ and $Y_2 = 21$.
- Display the graphs in an appropriate viewing window, such as $[0, 4.7] \times [0, 30]$.
- Use the **intersect** feature of your calculator to estimate the x-coordinate of the point of intersection of the two graphs, as shown in the accompanying figures.

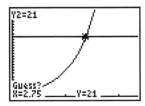 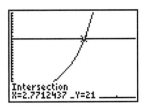

Solving logarithmic equations: An equation that contains logarithms is solved by eliminating the logarithms, either by writing the equation in exponential form or by consolidating the logarithms on both sides of the equation. The new equation has the form $\log_b N = \log_b M$, which implies $N = M$.

KEY EXAMPLE

Solve for x: $\log_3(7x + 4) - \log_3 2 = 2 \log_3 x$

Solution: Use the reverse of the quotient law of logarithms to express the left side of the given equation as the logarithm of a single term. On the right side of the original equation use the reverse of the power law of logarithms. Then

$$\log_3\left(\frac{7x + 4}{2}\right) = \log_3 x^2.$$

This equation has the form $\log_b N = \log_b M$, which implies $N = M$ where

$$N = \frac{7x + 4}{2} \quad \text{and} \quad M = x^2.$$

Thus,

$$\frac{7x + 4}{2} = x^2.$$

Multiplying each side of the equation by 2 and writing the resulting quadratic equation in standard form gives

$$2x^2 - 7x - 4 = 0,$$

whose roots are $x = -\frac{1}{2}$ and $x = 4$. For $x = -\frac{1}{2}$ the right side of the original equation evaluates to the logarithm of a negative quantity, which is not defined. Hence, the only root of the original equation is $x = 4$.

KEY EXAMPLE

Solve for x: $\log_4(x - 3) + \log_4(x + 3) = 2$

Solution: Rewrite the given equation as

$$\log_4[(x - 3)(x + 3)] = 2, \text{ so } (x - 3)(x + 3) = 4^2.$$

Simplifying this equation gives $x^2 = 25$ so $x = -5$ or $x = 5$. Reject $x = -5$ since substituting this value for x in the original equation produces the logarithm of a negative quantity, which is not defined. Thus, $x = 5$ is the only solution of the original equation.

KEY EXAMPLE

Solve for y in terms of e: $\ln 3y - \ln(y - 4) = 2$

Solution: Applying the reverse of the quotient law of logarithms on the left side of the given equation gives

$$\ln\left(\frac{3y}{y-4}\right) = 2 \text{ , so } \frac{3y}{y-4} = e^2.$$

Then

$$3y = e^2y - 4e^2, \text{ so } y = \frac{4e^2}{e^2 - 3}.$$

Key 53   Exponential growth and decay

OVERVIEW   *The growth patterns of many types of biological and physical quantities closely approximate an exponential curve.*

Compound interest: If an amount of money A_0 is invested at a fixed annual rate r and interest is compounded n times each year, then the amount of money in t years, denoted by $A(t)$, is given by the equation

$$A(t) = A_0\left(1 + \frac{r}{n}\right)^{nt}.$$

Example:

If \$1,200 is invested at an annual rate of 8% for 5 years and interest is compounded quarterly each year, then $A_0 = 1200$, $r = 0.08$, $t = 5$, and $n = 4$. Thus,

$$A(5) = 1200(1 + 0.02)^{5(4)} = 1200(1.02)^{20} = 1783.14.$$

In 5 years the investment grows from \$1,200 to \$1,783.14.

A model of natural exponential growth: An equation of the form

$$A(t) = A_o e^{kt}$$

predicts the growth of a subtance where A_0 is the amount of the substance present at $t = 0$ and $A(t)$ is the amount that will be present at time t. The number k is a constant whose value depends on the specific nature of the growth process. For example, if an amount of money A_0 is invested at a fixed annual rate r and interest is *continuously* compounded for t years, then $A(t) = A_0\, e^{\,rt}$.

Example:

If \$1,200 is invested at an annual rate of 8% for 5 years and interest is compounded continuously, then

$$A(5) = 1200e^{(0.08)5} = 1200e^{0.40} = 1790.19.$$

Thus, in 5 years the investment grows from \$1,200 to \$1,790.19.

Model of natural exponential decay: If k is negative, an equation of the form $A(t) = A_0 e^{kt}$ models exponential decay since $A(t)$ *decreases* over time. The disintegration of radioactive elements follows this type of law.

KEY EXAMPLE

If \$2,750 is invested at an annual rate of 9%, find the number of years it will take for the investment to double if interest is continuously compounded.

Solution: Use the formula $A(t) = A_0 e^{rt}$, where $A_0 = 2750$, $r = 0.09$, and $A(t) = 2(2750) = 5500$. Thus,

$$5500 = 2750 e^{0.09t}.$$

Dividing both sides of the equation by 2750 gives

$$2 = e^{0.09t}.$$

In logarithmic form the equation is

$$0.09t = \ln 2.$$

Thus,

$$t = \frac{\ln 2}{0.09}.$$

Use a calculator: $\ln 2$ is approximately 0.693, so

$$t = 0.693 \div 0.09 = 7.7.$$

The investment will double in approximately 7.7 years.

KEY EXAMPLE

The number of bacteria in a certain culture is 900. If the bacteria grow in the culture according to the law $A(t) = A_0 e^{0.5t}$, what will be the number of bacteria in the culture 2 hours from now?

Solution: At $t = 0$, $A_0 = 900$. Let $t = 2$. Then

$$A(2) = 900 e^{0.5(2)} = 2446.45.$$

Two hours from now the number of bacteria in the culture will be approximately 2446.

KEY EXAMPLE

A radioactive element decays exponentially according to the law $A(t) = A_0 e^{kt}$. After 200 days a 25-milligram sample of this element weighs 6 milligrams. Find the value of the decay constant k.

Solution: At $t = 0$, $A_0 = 25$. Let $t = 200$ and $A(200) = 6$. Then,

$$6 = 25e^{k(200)},$$

$$\text{so } e^{200k} = \frac{6}{25}.$$

In logarithmic form,

$$200k = \ln\left(\frac{6}{25}\right).$$

Thus,

$$k = \frac{\ln 6 - \ln 25}{200} = \frac{1.792 - 3.219}{200} = -0.007135$$

KEY EXAMPLE

The half-life of a radioactive substance is the amount of time it takes for one-half of the substance to disintegrate. If a radioactive substance decays according to the law $A(t) = A_0 e^{kt}$, where t is measured in days, find its half-life to the nearest day if $k = -0.025$.

Solution: In the formula $A(t) = A_0 e^{kt}$, let $A(t) = 0.5A_0$, $k = -0.025$, and then solve for t. Thus,

$$0.5A_0 = A_0 e^{(-0.025)t} \quad \text{or} \quad 0.5 = e^{-0.025t}.$$

In logarithmic form the equation is

$$-0.025t = \ln 0.5.$$

Use a calculator: -0.693 is an approximation for $\ln 0.5$. Thus,

$$t = -0.693 \div (-0.025) = 27.72.$$

The half-life to the nearest day is 28.

KEY EXAMPLE

Because the half-life of a radioactive element is the amount of time required for one-half of an initial amount of the element to decay, the process can be modeled by the function

$$A(t) = A_0 \left(0.5\right)^{\frac{t}{H}},$$

where H is the half-life of the substance, and $A(t)$ is the amount of the original sample, A_0, that remains after t units of time. If the half-life of radioactive strontium-90 is known to be 29 years, find the number of years that have elapsed if 30% of a 250-gram mass of strontium-90 remains.

Solution: Use the equation $A(t) = 250(0.5)^{\frac{t}{29}}$, where $A_0 = 250$ and $H = 29$. Let x represent the number of years that have elapsed if 30% of a 250-gram mass of strontium-90 remains. Thus:

$$A(x) = 250(0.5)^{\frac{x}{29}}$$

$$0.30\left(\cancel{250}\right) = \cancel{250}(0.5)^{\frac{x}{29}}$$

$$0.30 = (0.5)^{\frac{x}{29}}$$

$$\frac{x}{29}\log 0.5 = \log 0.30$$

$$x = \frac{29 \times \log 0.30}{\log 0.5} \approx 50.4 \text{ years}$$

Keep in mind that:
- The equation $0.3 = (0.5)^{\frac{t}{29}}$ can be solved by graphing $Y_1 = (0.5)^{\frac{x}{29}}$ and $Y_2 = 0.3$ in an appropriate viewing window, such as $[0, 75.2] \times [0, 0.5]$, and then using the **intersect** feature to estimate the x-coordinate of the point of intersection of the two graphs.

- In the equation $A(t) = A_0 (0.5)^{\frac{t}{H}}$, the numerator and the denominator of the exponent must be expressed in the same unit of time.

Theme 9 CONIC SECTIONS AND
THEIR EQUATIONS

A *circle, ellipse, parabola*, and *hyperbola* are plane curves, called **conic sections**, that can be formed by cutting a double cone with a plane at various angles, as shown in the figures below. A **circle** is formed when the plane intersects one cone and is perpendicular to the axis; if the plane is not perpendicular to the axis, an **ellipse** is formed. A **parabola** is formed when the plane intersects one cone and is parallel to an edge of the cone. A **hyperbola** is formed when a plane intersects both cones.

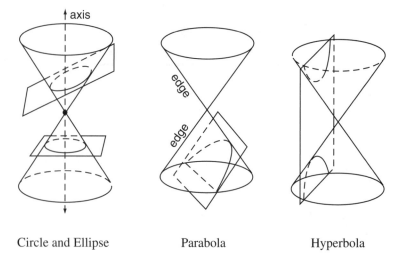

Circle and Ellipse   Parabola   Hyperbola

Conic Sections

Conic sections arise in diverse areas of applied mathematics. They can be used to help describe the paths of projectiles, planetary motion, and the shapes of light-reflecting surfaces. Each of the conic-section curves can be defined geometrically and its equation derived using the methods of analytic (coordinate) geometry.

Key 54 The circle

OVERVIEW   *A circle is defined geometrically in terms of a fixed point in a plane and a number.*

Circle: A circle is the set of all points P that are a fixed distance r, called the **radius**, from a given point (h, k), called the **center**. In Figure 9.1, $P(x, y)$ is on the circle if and only if its distance from (h, k) equals r:

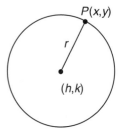

$$\sqrt{(x-h)^2 + (y-k)^2} = r.$$

Squaring both sides of this equation gives the standard form of an equation of a circle,

Figure 9.1   Circle

$$(x-h)^2 + (y-k)^2 = r^2.$$

Thus, the graph of the equation $x^2 + y^2 = r^2$ is a circle whose radius is r and whose center is at $(0, 0)$.

KEY EXAMPLE

Write an equation of a circle with the following:

(a) A radius length of 5 and center at $(-1, 4)$.
(b) A diameter whose endpoints are $A(-7, 5)$ and $B(1, 5)$.

Solution: (a) In the equation

$$(x-h)^2 + (y-k)^2 = r^2,$$

let $(h, k) = (-1, 4)$ and $r = 5$ so

$$[x-(-1)]^2 + (y-4)^2 = 5^2.$$

Simplifying gives

$$(x+1)^2 + (y-4)^2 = 25$$

as an equation of a circle with radius 5 and center $(-1, 4)$.

(b) The midpoint of $\overline{AB}$ is $(-3, 5)$ which is the center of the circle. Using the distance formula, you can verify that $AB = 8$, so the radius is 4. The equation of a circle with a radius length of 4 and center at $(-3, 5)$ is

$$(x+3)^2 + (y-5)^2 = 16.$$

Key 55 The ellipse

OVERVIEW *Given two fixed points* F₁ *and* F₂ *and a positive number* k, *an **ellipse** is traced out by connecting the set of all points* P *in the plane that satisfy the condition that* PF₁ + PF₂ = k *(see Figure 9.2).*

Definition of ellipse: An ellipse is the set of all points in the plane such that the sum of the distances from two fixed points, F_1 and F_2, is constant. The two fixed points are called the **foci** (plural of focus).

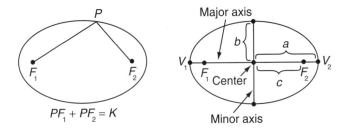

Figure 9.2   Ellipse Definition   **Figure 9.3**   Ellipse Terms

Terms associated with an ellipse:

*   **Center:** the midpoint of the segment whose endpoints are the two foci. The distance from the center to either focus is represented by *c*.
*   **Vertices:** the points V_1 and V_2 on the ellipse that are the endpoints of the line segment that contains the foci. The distance from the center to either of the two vertices is denoted by *a*.
*   **Axes of symmetry:** An ellipse has two axes of symmetry that pass through its center. The **major axis** is the line segment whose endpoints are V_1 and V_2. The **minor axis** is the line segment that is perpendicular to the major axis and whose endpoints are points on the ellipse each of which is *b* units from the center where $b^2 = a^2 - c^2$ ($a > b > 0$). Since the length of the major axis is $2a$ and the length of the minor axis is $2b$, the major axis is always the longer of the two axes of symmetry. If the axes of symmetry have the same length, then the ellipse is a circle.

Ellipse equations with center at (0, 0): We can compare ellipses with horizontal and vertical major axes in Figures 9.4 and 9.5. Each ellipse is labeled with its equation. The foci are located at $F(\pm c, 0)$ for a horizontal ellipse and at $F(0, \pm c)$ for a vertical ellipse.

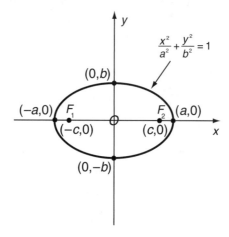

Figure 9.4 Horizontal Ellipse

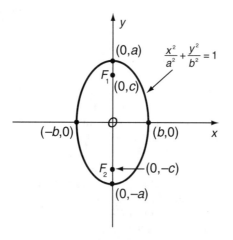

Figure 9.5 Vertical Ellipse

Ellipse equations with center at (h, k): The standard form of the equation of an ellipse whose center is at (h, k) is given in Table 9.1 where $b^2 = a^2 - c^2$ $(a > b > 0)$ or, equivalently, $c = \sqrt{a^2 - b^2}$.

Table 9.1 Ellipse Equations Where $c = \sqrt{a^2 - b^2}$ and $a > b > 0$

Major Axis	Equation	Foci	Vertices
Horizontal	$\dfrac{(x-h)^2}{a^2} + \dfrac{(y-k)^2}{b^2} = 1$	$F(h \pm c, k)$	$V(h \pm a, k)$
Vertical	$\dfrac{(x-h)^2}{b^2} + \dfrac{(y-k)^2}{a^2} = 1$	$F(h, k \pm c)$	$V(h, k \pm a)$

KEY EXAMPLE

Find the coordinates of the foci and vertices of an ellipse whose equation is $\dfrac{x^2}{16} + \dfrac{y^2}{9} = 1$.

Solution: Since the x^2-term has the larger denominator, the major axis is horizontal. Let $a^2 = 16$ and $b^2 = 9$. Then $a = 4$, $b = 3$, and $c = \sqrt{a^2 - b^2} = \sqrt{7}$. Since the center is at the origin and the major axis coincides with the x-axis, the foci are at $F(\pm c, 0) = F_1(-\sqrt{7}, 0)$ and $F_2(\sqrt{7}, 0)$ and the vertices are at $V(\pm a, 0) = V_1(-4, 0)$ and $V_2(4, 0)$.

KEY EXAMPLE

Find the coordinates of the foci and vertices of an ellipse whose equation is $25x^2 + 4y^2 = 100$.

Solution: Put the given equation in standard form by dividing each term of the equation by 100. Then,

$$\frac{x^2}{4} + \frac{y^2}{25} = 1.$$

Since the y^2-term has the larger denominator and the center is at the origin, the major axis is vertical and lies along the y-axis. Let $a^2 = 25$ and $b^2 = 4$. Then $a = 5$, $b = 2$, and $c = \sqrt{a^2 - b^2} = \sqrt{21}$. Hence, the foci are at $F(0, \pm c) = F_1(0, -\sqrt{21})$ and $F_2(0, \sqrt{21})$ and the vertices are at $V(0, \pm a) = V_1(0, -5)$ and $V_2(0, 5)$.

KEY EXAMPLE

For the ellipse whose foci are (–1, 2) and (7, 2) and whose minor axis has a length of 6, find the center, the vertices, and the standard form of its equation.

Solution: Since the foci have the same *y*-coordinate, the major axis is horizontal. The center (h, k) of the ellipse is the midpoint of the segment whose endpoints are the foci. Hence,

$$(h,k) = \left(\frac{-1+7}{2}, \frac{2+2}{2} \right) = (3,2).$$

Since the distance from either focus to (3, 2) is 4, $c = 4$. The length of the vertical minor axis is given as 6, so $2b = 6$ and $b = 3$. Thus,

$$a = \sqrt{c^2 + b^2} = \sqrt{4^2 + 3^2} = 5.$$

The vertices are at $V(h \pm a, k) = V(3 \pm 5, 2) = V_1(-2, 2)$ and $V_2(8, 2)$. Let $(h, k) = (3, 2)$, $a = 5$, and $b = 3$ in the standard form of the equation of a horizontal ellipse; then

$$\frac{(x-3)^2}{25} + \frac{(y-2)^2}{9} = 1.$$

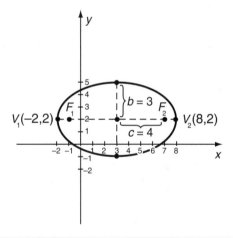

Key 56 The hyperbola

OVERVIEW *An ellipse is formed by connecting the set of all points* P *such that the **sum** of the distances from the foci to* P *is constant. If the **difference** of these distances is taken, a **hyperbola** is formed.*

Definition of hyperbola: A hyperbola is the set of all points P in the plane such that the difference of the distances from P to two fixed points, called **foci**, is constant. As shown in Figure 9.6, a hyperbola consists of two nonintersecting curves called **branches**. The branches resemble a pair of parabolas that open in opposite directions.

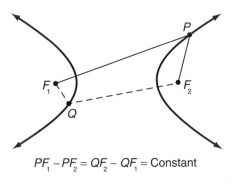

$$PF_1 - PF_2 = QF_2 - QF_1 = \text{Constant}$$

Figure 9.6 Hyperbola Definition

Terms associated with a hyperbola shown in Figure 9.7:

- **Center:** the midpoint of the segment whose endpoints are the foci. As with an ellipse, c represents the distance from the center to a focus point.
- **Vertices:** the points V_1 and V_2 at which the line that contains the foci intersects the branches of the hyperbola. The distance from the center to either of the two vertices is denoted by a. Unlike the case of an ellipse, $c > a$ for a hyperbola.

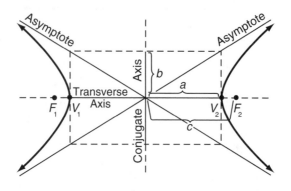

Figure 9.7 Hyperbola Terms

- **Axes of symmetry:** Like an ellipse, a hyperbola has two axes of symmetry that pass through its center. The **transverse axis** is the line segment whose endpoints are V_1 and V_2. The **conjugate axis** is the line segment that is perpendicular to the transverse axis and whose endpoints are each b units from the center where a, b, and c are related by the equation $b^2 = c^2 - a^2$ $(c > a > 0)$. The length of the transverse axis is $2a$, and the conjugate axis measures $2b$. Unlike the case of an ellipse, either axis of symmetry may be longer than the other, or both axes may have the same length.
- **Hyperbola asymptotes:** two lines that the branches of the curve approach as they depart from the center. The asymptotes contain the diagonals of the rectangle that has the same center as the hyperbola and whose adjacent sides measure $2a$ and $2b$.

Hyperbola equations with center at (0, 0): Figures 9.8 and 9.9 show hyperbolas with horizontal and vertical transverse axes. Each hyperbola is labeled with its equation. The foci are located at $F(\pm c, 0)$ for a horizontal hyperbola and at $F(0, \pm c)$ for a vertical hyperbola. The equations of the asymptotes are $y = \pm \dfrac{b}{a} x$ for a horizontal hyperbola, and $y = \pm \dfrac{a}{b} x$ for a vertical hyperbola.

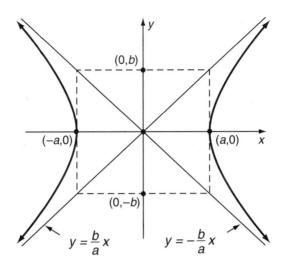

Figure 9.8 Horizontal Hyperbola

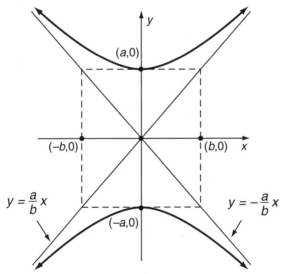

Figure 9.9 Vertical Hyperbola

Hyperbola equations with center at (h, k): The standard forms of the equations of a hyperbola whose center is at (h, k) are given in Table 9.2, where $b^2 = c^2 - a^2$ $(c > a > 0)$ or, equivalently, $c = \sqrt{a^2 + b^2}$.

Table 9.2 Hyperbola Equations

	Horizontal Transverse Axis	**Vertical Transverse Axis**
Equation	$\dfrac{(x-h)^2}{a^2} - \dfrac{(y-k)^2}{b^2} = 1$	$\dfrac{(y-k)^2}{a^2} - \dfrac{(x-h)^2}{b^2} = 1$
Foci	$F(h \pm c, k)$	$F(h, k \pm c)$
Vertices	$V(h \pm a, k)$	$V(h, k \pm a)$
Asymptotes	$y - k = \pm\dfrac{b}{a}(x - h)$	$y - k = \pm\dfrac{a}{b}(x - h)$

KEY EXAMPLE

What is an equation of a hyperbola whose vertices are at $(\pm 12, 0)$ and whose foci are at $(\pm 13, 0)$?

Solution: Since the vertices lie on the x-axis and are the endpoints of a segment whose midpoint is the origin, the hyperbola has a horizontal transverse axis and its center at $(h, k) = (0, 0)$. The vertices of this type of hyperbola are at $V(\pm a, 0) = V(\pm 12, 0)$, and the foci are at $F(\pm c, 0) = F(\pm 13, 0)$. Since $a = 12$ and $c = 13$,

$$b = \sqrt{c^2 - a^2} = \sqrt{13^2 - 12^2} = 5.$$

Thus, the standard equation of this horizontal hyperbola is

$$\frac{x^2}{144} - \frac{y^2}{25} = -1.$$

KEY EXAMPLE

What are the coordinates of the center, vertices, and foci of the hyperbola whose equation is $\dfrac{(y-1)^2}{9} - \dfrac{(x+4)^2}{7} = 1$?

Solution: The equation has the form

$$\frac{(y-k)^2}{a^2} - \frac{(x-h)^2}{b^2} = 1$$

where $h = -4$, $k = 1$, $a^2 = 9$, $b^2 = 7$, and $c = \sqrt{a^2 + b^2} = \sqrt{25} = 5$. Thus, the hyperbola has a vertical transverse axis with center $= (h, k) = (-4, 1)$, vertices $= V(h, k \pm a) = V(-4, 1 \pm 3) = V_1(-4, -2)$ and $V_2(-4, 4)$, and foci $= F(h, k \pm c) = F(-4, 1 \pm 5) = F_1(-4, -4)$ and $F_2(-4, 6)$.

KEY EXAMPLE

Find the standard equation of the hyperbola whose foci are at $(-1, -3)$ and $(-1, 7)$ and whose conjugate axis has a length of 8. Sketch its graph.

Solution: First determine the equation of the hyperbola. The foci are on the same vertical line since they have the same x-coordinate. Hence, the hyperbola has a transverse axis whose length $= 7 - (-3) = 10$. The center (h, k) is the midpoint of the segment whose endpoints are the foci. Hence

$$(h,k) = \left(\frac{-1 + (-1)}{2}, \frac{-3 + 7}{2} \right) = (-1, 2).$$

The distance c from the center to either of the two foci is 5. Since the length of the conjugate axis is 8, $2b = 8$, so $b = 4$. Thus,

$$a = \sqrt{c^2 - b^2} = \sqrt{5^2 - 4^2} = 3.$$

Since $h = -1$, $k = 2$, $a^2 = 9$, and $b^2 = 16$, the standard equation of this vertical hyperbola is

$$\frac{(y-2)^2}{9} - \frac{(x+1)^2}{16} = 1.$$

Graph the center $(h, k) = (-1, 2)$ and the vertices $V(h, k \pm a) = V(-1, 2 \pm 3) = V_1(-1, -1)$ and $V_2(-1, 5)$. Draw a rectangle with base $2b = 8$ and height $2a = 6$ such that the vertices are the midpoints of a pair of opposite sides of the rectangle. Then draw the asymptotes through the diagonals of the rectangle. Sketch the hyperbola, using the asymptotes as guideposts, as the branches of the curve move away from the vertices.

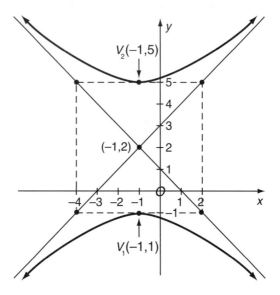

Rectangular hyperbolas: The graph of the equation $xy = k$ ($k \neq 0$) is a special type of hyperbola, called a **rectangular** or **equilateral hyperbola**. The foci of this type of hyperbola lie on the line $y = x$.

Key 57 The parabola

OVERVIEW *Given a line and a fixed point not on the line, a **parabola** is traced out by connecting the set of all points Q whose distances from the line and from the fixed point are the same.*

Parabola: A parabola is the set of all points Q in the plane such that the distance from Q to a fixed point F, called the **focus**, is equal to the distance from Q to a given line ℓ, called the **directrix**. See Figure 9.10.

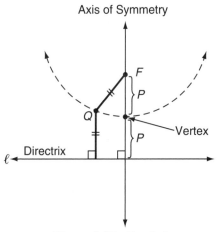

Axis of Symmetry

Figure 9.10 Parabola

Axis of symmetry: The axis of symmetry is the line that contains the focus and is perpendicular to the directrix.

Vertex (turning point): The point V at which the axis of symmetry intersects the parabola is the vertex. The directrix and the focus are on opposite sides of the vertex and are the same distance from it. The directed distance from the vertex to either the focus or the directrix is denoted by p.

Parabola equations: The standard form of the equation of a parabola whose vertex is (h, k) is given in Table 9.3.

Table 9.3   Standard Form of Parabola Equations

Parabola Equation	$p > 0$	$p < 0$
$(x - h)^2 = 4p(y - k)$ where • axis of symmetry: $x = h$ • focus: $F(h, k + p)$ • directrix: $y = k - p$		
$(y - k)^2 = 4p(x - h)$ where • axis of symmetry: $y = k$ • focus: $F(h + p, k)$ • directrix: $x = h - p$		

KEY EXAMPLE

Write an equation of the parabola whose vertex is at $V(0, 0)$ and whose focus is at $F(0, 3)$. Also, write equations for the axis of symmetry and for the directrix.

Solution: The vertex and focus lie on the same vertical line (the y-axis) so the standard equation of the parabola has the form

$$(x - h)^2 = 4p(y - k)$$

where $(h, k) = (0, 0)$. Since the focus is located 3 units *above* the vertex, $p = 3$. Hence, an equation of this vertical parabola is

$$x^2 = 12y.$$

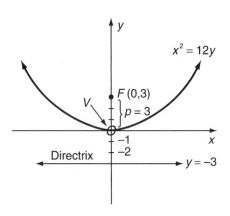

An equation of the axis of symmetry is

$$x = h = 0.$$

The directrix is $|p|$ units on the opposite side of the vertex and is perpendicular to the axis of symmetry. Since $p = 3$, an equation of the directrix is

$$y = k - p = 0 - 3 = -3.$$

KEY EXAMPLE

Determine the coordinates of the vertex and the focus of the parabola whose equation is $y^2 - 6y + 8x + 17 = 0$. Give equations for the axis of symmetry and for the directrix.

Solution: The given equation can be put into the form

$$(y - k)^2 = 4p(x - h)$$

by using the method of completing the square. Isolating terms involving y in the original equation gives

$$y^2 - 6y = -8x - 17.$$

Complete the square by adding the square of one-half of -6 to both sides of the equation. Thus,

$$y^2 - 6y + 9 = -8x - 17 + 9$$
$$(y - 3)^2 = -8x - 8$$
$$(y - 3)^2 = -8(x + 1)$$

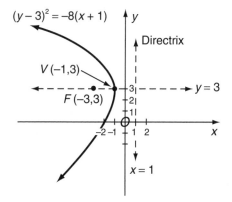

The equation $(y - 3)^2 = -8(x + 1)$ has the form $(y - k)^2 = 4p(x - h)$ where $h = -1$, $k = 3$, and $4p = -8$ so $p = -2$. Thus, the vertex $V(h, k) = (-1, 3)$. Since p is negative, the parabola opens to the left and the focus is 2 units to left of $(-1, 3)$. Hence, the focus is at

$$F(h + p, k) = F(-1 + (-2), 3) = F(-3, 3).$$

An equation of the horizontal axis of symmetry is

$$y = k = 3.$$

An equation of the vertical directrix is

$$x = h - p = -1 - (-2) = 1.$$

Key 58 The general equation

for conic sections

OVERVIEW   *Different conic sections can be described by the same general equation.*

Conic section equations with center at (0, 0): The curve traced out by the equation $Ax^2 + By^2 = C$ ($C \neq 0$) depends on the values of the constants A, B, and C.

1. The graph is a **circle** if A and B are equal and have the same sign as C.
2. The graph is an **ellipse** if A and B are unequal and have the same sign as C.
3. The graph is a **hyperbola** if A and B have different signs.

Conic section equations with center at (h, k): If the standard equations for a circle, ellipse, hyperbola, and parabola are expanded and simplified, then each of the resulting equations will have the form $Ax^2 + By^2 + Cx + Dy + E = 0$ where A, B, C, D, and E are real numbers with at least one of coefficients A and B not 0.

Degenerate conics: For some special sets of values of the coefficients of the equation $Ax^2 + By^2 + Cx + Dy + E = 0$, the graph of the equation is a *degenerate conic*—a point, line, or pair of lines. For example, the graph of $x^2 + y^2 = 0$ is the point (0, 0), and the graph of $x^2 - y^2 = 0$ is a pair of intersecting lines whose equations are $y = x$ and $y = -x$. For some numerical coefficients, the general equation may not have any real solution pairs so the graph is the empty set, as in $x^2 + y^2 + 16 = 0$.

Identifying a conic section from its equation: If the graph of $Ax^2 + By^2 + Cx + Dy + E = 0$ is not a degenerate conic, then its graph can be identified by comparing A and B:

- If exactly one of A or B is 0, the graph is a parabola.
- If $A = B$, the graph is a circle.
- If $A \neq B$ and A and B have the same sign, the graph is an ellipse.
- If A and B have opposite signs, the graph is a hyperbola.

KEY EXAMPLE

Name the conic section whose equation is $x^2 - 9y^2 - 10x - 54y - 47 = 0$ and determine the coordinates of its center, vertices, and foci.

Solution: The given equation has the form $Ax^2 + By^2 + Dx + Ey + F = 0$ where A and B have opposite signs. Hence, its graph is a hyperbola.

To find its center, put the equation into standard form by completing the square for x and for y. Grouping like variable terms together gives

$$(x^2 - 10x + \text{?}) - 9(y^2 + 6y + \text{?}) = 47.$$

Completing the square for each quadratic polynomial yields

$$(x^2 - 10x + 25) - 9(y^2 + 6y + 9) = 47 + 25 - 81$$
$$(x - 5)^2 - 9(y + 3)^2 = -9$$

To put this equation in standard form, divide each member of the equation by –9. Thus,

$$\frac{(y+3)^2}{1} - \frac{(x-5)^2}{9} = 1.$$

The derived hyperbola equation has the form

$$\frac{(y-k)^2}{a^2} - \frac{(x-h)^2}{b^2} = 1,$$

in which $h = 5$, $k = -3$, $a^2 = 1$, $b^2 = 9$, and $c = \sqrt{a^2 + b^2} = \sqrt{1+9} = \sqrt{10}$. The graph is a vertical hyperbola with center at $(h, k) = (5, -3)$. The vertices are at $(h, k \pm a) = (5, -3 \pm 1) = V_1(5, -2)$ and $V_2(5, -4)$, and the foci are at $(h, k \pm c) = F_1(5, -3 - \sqrt{10})$ and $F_2(5, -3 + \sqrt{10})$.

Revisiting the general equation of conics: If the foci of a conic section do not lie on a line that is parallel to a coordinate axis, the general equation of that conic is

$$Ax^2 + Bxy + Cy^2 + Dx + Ey + F = 0,$$

where $B \neq 0$. If $B = 0$, the cross-product term vanishes and the general equation simplifies to

$$Ax^2 + Cy^2 + Dx + Ey + F = 0,$$

which describes a conic whose focal axis is parallel to a coordinate axis.

The discriminant test for classifying conics: The **discriminant** of the equation $Ax^2 + Bxy + Cy^2 + Dx + Ey + F = 0$ is the quantity $B^2 - 4AC$. Assume that A, B, and C are not all 0 and that degenerate cases are excluded. Then:

- If $B^2 - 4AC < 0$, the graph is an ellipse or a circle.
- If $B^2 - 4AC = 0$, the graph is a parabola.
- If $B^2 - 4AC > 0$, the graph is a hyperbola.

Key 59 Solving nonlinear systems

OVERVIEW *A **nonlinear system** is a system of equations that contains at least one equation whose graph is not a line. Some nonlinear systems can be solved with the same methods used to solve two-equation linear systems.*

Real solutions of a nonlinear system: The points at which the graphs of the equations of the system intersect, if any, represent real solutions. For some simple types of nonlinear systems these points may be determined by using one of the following methods:

- Graph the equations on the same set of axes and note the points where the graphs intersect. Graphing calculators can be helpful in finding or approximating these points.
- Express one of the equations in terms of a single variable that then can be solved using routine methods. Algebraic methods produce all possible solutions, real and nonreal. If a nonlinear system has no real solutions, then the graphs of its equations do not intersect.

Solving a linear-quadratic system: Figure 9.11 shows the possible number of points at which a line and a parabola may intersect. In general, the solution set of a system of one first-degree equation and one second-degree equation consists of, at most, two ordered pairs.

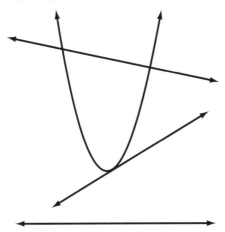

Figure 9.11 Possible Number of Points at Which
a Line and Parabola May Intersect

Example 1:   To solve
$$(x - 3)^2 + (y + 2)^2 = 10$$
$$x + y = 3$$

solve the linear equation for x: $x = 3 - y$. Substituting this expression in the first equation gives

$$(3 - y - 3)^2 + (y + 2)^2 = 10$$

which simplifies to

$$y^2 + 2y - 3 = 0.$$

The roots of this equation are $y = -3$ and $y = 1$. The corresponding values for x are obtained by substituting each value of y in $x + y = 3$ and then solving for x. Thus, the line $x + y = 3$ intersects the circle $(x - 3)^2 + (y + 2)^2 = 10$ in two points: $(6, -3)$ and $(2, 1)$.

Example 2:   To solve

$$xy + 3y = 7$$
$$2xy + y^2 = 4$$

multiply the first equation by -2 and then add the result to the second equation. This gives $y^2 - 6y = -10$ or in standard form

$$y^2 - 6y + 10 = 0.$$

Use the quadratic formula. Then,

$$y = \frac{6 \pm \sqrt{36 - 4(1)(10)}}{2} = 3 \pm i.$$

The nonreal solutions indicate that the curves defined by the given pair of equations do not intersect.

Solving a system of two quadratic equations: The solution set of a system of two quadratic equations contains, at most, four ordered pairs.

Example 3:   To solve

$$x^2 + y^2 = 13$$
$$xy = 6$$

solve the second equation for y: $y = \dfrac{6}{x}$. Substituting this expression in the first equation yields

$$x^2 + \left(\frac{6}{x}\right)^2 = 13 \quad \text{or} \quad x^2 + \frac{36}{x^2} = 13.$$

Clearing fractions by multiplying each term by x^2 leads to

$$x^4 - 13x^2 + 36 = 0.$$

This equation can be factored as

$$(x^2 - 9)(x^2 - 4) = 0.$$

If $x^2 - 9 = 0$, then $x = -3$ or $x = +3$. If $x^2 - 4 = 0$, then $x = -2$ or $x = +2$. The corresponding values for y are obtained by substituting each value of x in $xy = 6$ and then solving for y. Thus, the rectangular hyperbola $xy = 6$ intersects the circle $x^2 + y^2 = 13$ in four points: $(-3, -2)$, $(3, 2)$, $(-2, -3)$, and $(2, 3)$.

KEY EXAMPLE

Solve the following system of equations:

$$2y^2 - 11x^2 = -1$$
$$5x^2 + 2xy = 3$$

Solution: If $5x^2 + 2xy = 3$,

$$y = \frac{3 - 5x^2}{2x}, \text{ so } y^2 = \left(\frac{3 - 5x^2}{2x}\right)^2 = \frac{9 - 30x^2 + 25x^4}{4x^2}.$$

Then $2y^2 - 11x^2 = -1$ becomes

$$2\left(\frac{9 - 30x^2 + 25x^4}{4x^2}\right) - 11x^2 = -1 \text{ or, equivalently, } 3x^4 - 28x^2 + 9 = 0.$$

Thus, $(3x^2 - 1)(x^2 - 9) = 0$.

- If $x^2 - 9 = 0$, then $x^2 = 9$, so $x = \pm 3$. Substituting 9 for x^2 in the first equation gives

$$2y^2 - 99 = -1, \text{ so } y^2 = 49 \quad \text{and} \quad y = \pm 7.$$

- If $3x^2 - 1 = 0$, then $x^2 = \frac{1}{3}$ and $x = \pm \frac{1}{\sqrt{3}}$. Substituting $\frac{1}{3}$ for x^2 in the first equation gives

$$2y^2 - \frac{11}{3} = -1, \text{ so } y^2 = \frac{4}{3} \quad \text{and} \quad y = \pm \frac{2}{\sqrt{3}}.$$

The four solutions are $(+3, -7)$, $(-3, +7)$, $\left(\frac{1}{\sqrt{3}}, \frac{2}{\sqrt{3}}\right)$, and $\left(-\frac{1}{\sqrt{3}}, -\frac{2}{\sqrt{3}}\right)$.

Theme 10 COUNTING METHODS

*A*n important application of mathematics is counting the number of ways in which the members of a set can be arranged in a certain order or can be selected as members of a subset. Given enough time, these arrangements or selections can be counted by listing the members for each possible arrangement or selection. More efficient counting methods make use of formulas for **permutations** (ordered arrangements of objects) and **combinations** (unordered subsets of objects).

Key 60   The basic principles of counting

OVERVIEW   *If we know the number of ways in which each of two events can occur, we can find the number of ways in which both events or one of the two events can happen.*

Fundamental rules of counting: Assume event *A* can occur in *a* ways and event *B* can occur in *b* ways.

- **Multiplication principle of counting:** Events *A* and *B* can both occur in $a \times b$ ways. This rule can be extended to more than two events.

   Example 1:   If Bill has 6 different shirts, 4 different ties, and 3 different sport jackets, then the number of different outfits he can choose, each of which consists of a shirt, tie, and sport jacket, is $6 \cdot 4 \cdot 3 = 72$.

   Example 2:   The total number of possible arrangements of the three letters *A, B,* and *C* can be determined in the following way: The first position can be filled by any of the *three* letters. Any of the remaining *two* letters can occupy the middle position. Once the first and second positions are determined, the *one* letter left can be put in the last position. By the multiplication principle of counting, the total number of arrangements is $3 \cdot 2 \cdot 1 = 6$. Similarly, four letters can be arranged in $4 \cdot 3 \cdot 2 \cdot 1 = 24$ ways, five letters can be arranged in $5 \cdot 4 \cdot 3 \cdot 2 \cdot 1 = 120$ ways, and so forth.

- **Addition principle of counting:** Event *A* or *B* can occur in $a + b$ ways, provided the two events are **mutually exclusive**. In other words, two mutually exclusive events cannot occur at the same time, just as we cannot obtain a head *and* a tail when a coin is flipped.

Key 61 Permutations

OVERVIEW *The number of different ways in which objects can be arranged in a row can be counted using a simple formula.*

Permutation: A permutation is an arrangement of objects in a definite order.

Example: The set of letters $\{A, B, C\}$ can be arranged in six different orders:

$$ABC, ACB, BAC, BCA, CAB, \text{ and } CBA.$$

Each three-letter arrangement represents a *permutation* of the three letters.

n **Factorial:** The product of consecutive positive integers from n down to 1 is called n factorial and is denoted by $\boldsymbol{n}!$. By definition, $0! = 1$.

Example: $5! = 5 \cdot 4 \cdot 3 \cdot 2 \cdot 1 = 120$.

$_nP_r$ **:** This symbol denotes the number of different ordered arrangements of r out of n objects, where $0 \le r \le n$. The symbol $_nP_r$ is read as "the number of permutations of n things taken r at a time." Thus, $_nP_n = n!$ is the number of permutations when all n objects are used in each arrangement.

Permutation formula: If r and n are positive integers with $r \le n$, then

$$_nP_r = \frac{n!}{(n-r)!} = \underbrace{n(n-1)(n-2)...(n-r+1)}_{\substack{product \ of \ the \\ r \ greatest \ factors \ of \ n}}.$$

If $r = n$, the formula simplifies to

$$_nP_n = n!$$

Example: The number of different three-digit numbers that can be formed using digits 1, 2, 3, 4, 5, and 6, assuming no digit is used more than once, is

$$_6P_3 = 6 \cdot 5 \cdot 4 = 120.$$

If the digits can be used more than once, then the number of different three-digit numbers that can be formed is, by the multiplication principle of counting, $6 \times 6 \times 6 = 216$.

Permutations with conditions attached: Sometimes arrangements are formed subject to special conditions. For example, suppose we want to know the number of three-digit numbers greater than 500 that can be formed using the digits 1, 2, 3, 4, 5, 6, and 7. The digit used for the first position of each arrangement must satisfy the condition that it is 5 or greater. There are three such digits (5, 6, and 7). Assuming no repetition of digits, the second position of each three-digit arrangement may be filled by any of the remaining six digits and the last position may be filled by any of the remaining five digits. By the multiplication principle of counting, the total number of three-digit numbers greater than 500 that can be formed is $3 \times 6 \times 5 = 90$. We could have also reasoned that the first position may be filled in $_3P_1$ ways and the remaining two positions filled in $_6P_2$ ways. The number of three-digit numbers greater than 500 is $_3P_1 \cdot _6P_2 = 90$. If repetition of digits is allowed, then $3 \times 7 \times 7 = 147$ three-digit numbers greater than 500 can be formed.

Addition principle and permutations: Permutations may need to be added to obtain the total number of possible arrangements.

Example: How many numbers greater than 400 and less than 9999 can be formed using the digits 0, 1, 3, 4, and 5 if no digit is used more than once? Since the leading digit must be either 4 or 5, there are $2 \cdot 4 \cdot 3 = 24$ three-digit numbers greater than 400 that can be formed. Every four-digit number is greater than 400. However, since 0 cannot be used for the first digit, $4 \cdot 4 \cdot 3 \cdot 2 = 96$ four-digit numbers can be formed. Thus, the number of three-digit or four-digit numbers greater than 400 that can be formed is $24 + 48 = 120$.

Circular permutations: The number of ways in which n different objects can be arranged in a circle is $(n - 1)!$

Example: Five people can be seated at a round table in $(5 - 1)! = 4! = 24$ different ways.

Permutations of objects with some identical: If in a set of n objects, a objects are identical, b objects are identical, c objects are identical, and so forth, then the number of different ways in which the n objects taken all at a time can be arranged is

$$\frac{n!}{a!\,b!\,c!\ldots}.$$

Example: The number of different ways four red flags, three blue flags, and one green flag can be arranged on a vertical flagpole is

$$\frac{(4+3+1)!}{4!\,3!\,1!} = \frac{8!}{4!\,3!} = \frac{8 \cdot 7 \cdot \overset{1}{\cancel{6}} \cdot 5 \cdot \overset{1}{\cancel{4!}}}{\cancel{4!} \cdot \cancel{3} \cdot \cancel{2} \cdot 1} = 280.$$

Key 62 Combinations

OVERVIEW *Although there are six **permutations** of the three letters* A, B, *and* C, *there is only one distinct set or* **combination** *of the three letters*—{A, B, C}.

$_nC_r$: This symbol denotes the number of subsets with r members that can be formed from a set with n members where $0 \leq r \leq n$. The symbol $_nC_r$ is read as "the number of combinations of n things taken r at a time."

Combination formula: If $0 \leq r \leq n$, then

$$_nC_r = \frac{_nP_r}{r!} = \frac{n!}{r!(n-r)!}.$$

Combinations vs. permutations: Permutations are *ordered* arrangements of objects while **combinations** are collections of objects in which only the identities of the objects matter. For example, suppose three people are chosen from a group of five people. If the people stand on a line, then their order is significant and the number of different ordered arrangements of three out of five people is $_5P_3 = 5 \cdot 4 \cdot 3 = 60$. If the three people are selected to serve on a committee, however, their names rather than their order are important, so the total number of three-member committees that can be formed is $_5C_3$, where

$$_5C_3 = \frac{5!}{3!(5-2)!} = \frac{5!}{3!2!} = \frac{5 \cdot \overset{2}{\cancel{4}} \cdot \overset{1}{\cancel{3!}}}{\cancel{3!} \cdot \cancel{2} \cdot 1} = 10.$$

Combinations for event *A AND* event *B*: Multiplying the number of combinations for event *A* by the number of combinations for event *B* gives the number of selections for the two events occurring together.

Example: There are six pens and seven books on a desk. If a student wants to take four pens (event *A*) *and* three books from the desk (event *B*), how many selections are possible? Four pens can be selected from six pens in $_6C_4$ ways. Three books can be selected from seven books in $_7C_3$ ways. Thus, the number of selections that include four pens *and* three books is $_6C_4 \cdot _7C_3 = 525$.

Combinations for event *A OR* event *B*: Adding the number of combinations for event *A* to the number of combinations for a mutually exclusive event *B* gives the number of selections for exactly one of the two events occurring.

Example: A jar contains four blue marbles and six green marbles. If three marbles are drawn without looking, in how many ways can the three marbles be selected so that they have the same color? The three marbles will have the same color if three blue marbles are selected (event *A*) *or* if three green marbles are selected (event *B*). Three blue marbles can be selected from four blue marbles in $_4C_3$ ways. Three green marbles can be selected from six green marbles in $_6C_3$ ways. Thus, the total number of ways in which three marbles having the same color can be selected is

$$_4C_3 + {_6C_3} = 4 + 20 = 24.$$

Combinatorial relationships: These formulas can often save time in evaluating combinations that have the indicated forms:

- $_nC_n = 1$
- $_nC_1 = n$
- $_nC_0 = 1$
- $_nC_k = {_nC_{n-k}}$ $(k \leq n)$

Combinations taken from 1 to *n*: The sum of the combinations of *n* different objects taken 1, 2, 3, . . . , *n* at a time is $2^n - 1$. For example, the total number of different subcommittees consisting of at least one person that can be selected from a group of five people is $2^5 - 1 = 31$.

KEY EXAMPLE

From a committee of seven Republicans and five Democrats a five-member subcommittee is to be selected. In how many different ways can the subcommitte be selected if *at least* three of its members must be Democrats?

Solution: The subcommitte will have *at least* three Democrats if it includes three, four, or five Democrats.

A subcommittee of 3 Democrats *and* 2 Republicans can be selected in $_5C_3 \cdot {_7C_2} = 210$ ways.
A subcommittee of 4 Democrats *and* 1 Republican can be selected in $_5C_4 \cdot {_7C_1} = 35$ ways.
A subcommittee of 5 Democrats *and* 0 Republican can be selected in $_5C_5 \cdot {_7C_0} = 1$ way.

Thus, there is a total of $210 + 35 + 1$ ways in which a five-member subcommittee that includes *at least* three Democrats can be selected.

Key 63   The binomial theorem

OVERVIEW   *Expanding a binomial to a positive integer power, as in*

$$(a + b)^5 = a^5 + 5a^4b + 10a^3b^2 + 10a^2b^3 + 5ab^4 + b^5$$

*can be accomplished using a formula called the **binomial theorem**.*

Binomial theorem: For any positive integer n, the expansion of $(a + b)^n$ is given by the formula

$$(a + b)^n = {}_nC_0a^n + {}_nC_1a^{n-1}b + {}_nC_2a^{n-2}b^2 + {}_nC_3{}^{n-3}b^3 + \cdots$$
$$\cdots + {}_nC_{n-1}a^1b^{n-1} + {}_nC_nb^n$$

where the numerical coefficients of the $n + 1$ terms are represented in combinatorial notation.

In the expansion formula for $(a + b)^n$:

- since ${}_nC_0 = {}_nC_n = 1$, the first term is a^n and the last term is b^n;
- the coefficient of the kth term is ${}_nC_{k-1}$;
- the sum of the exponents in each term is n;
- in successive terms after the first, the power of a diminishes by 1, while the power of b increases by 1.

Pascal's triangle: Writing the values of ${}_nC_r$ in a triangular array in which the rows correspond to successive values of n, starting with $n = 0$, as shown in the accompanying figure, forms a pattern of numbers called *Pascal's triangle*. Pascal's triangle is named after the French mathematician Blaise Pascal (1623–1662).

row 0:				${}_0C_0$					1	
row 1:			${}_1C_0$	${}_1C_1$				1	1	
row 2:			${}_2C_0$ ${}_2C_1$ ${}_2C_2$				1	2	1	
row 3:		${}_3C_0$ ${}_3C_1$ ${}_3C_2$ ${}_3C_3$				1	3	3	1	
row 4:	${}_4C_0$ ${}_4C_1$ ${}_4C_2$ ${}_4C_3$ ${}_4C_4$				1	4	6	4	1	
row 5:	${}_5C_0$ ${}_5C_1$ ${}_5C_2$ ${}_5C_3$ ${}_5C_5$			1	5	10	10	5	1	

.
.
.

· · ·   · · ·

Pascal's Triangle

In Pascal's triangle, each number after 1 is the sum of the two numbers directly above it. If you know this fact and the symmetry pattern that results from it, it is easy to create Pascal's triangle. This pattern allows you to find quickly the value of $_nC_r$ for different values of n and r, where n is the row number and $r = 0$ corresponds to the first entry on each row.

KEY EXAMPLE

Write the expansion of $(x - 3y)^4$.

Solution: Use the binomial theorem to expand $(a + b)^n$, where $a = x$, $b = -3y$, and $n = 4$:

$$(x - 3y)^4 = {}_4C_0\, x^4 + {}_4C_1\, x^3(-3y)^1 + {}_4C_2\, x^2(-3y)^2 + {}_4C_3\, x^1(-3y)^3 + {}_4C_4\, x^0(-3y)^4.$$

Evaluate the combinations either by using a calculator or by copying the values from row 4 of Pascal's triangle shown in the accompanying figure:

$$
\begin{aligned}
&= \;\; 1x^4 \;\;\;\; + \;\; 4x^3(-3y) \;\;\;\; + \;\; 6x^2(-3y)^2 \;\;\;\; + \;\; 4x(-3y)^3 \;\;\;\; + \;\; 1(-3y)^4 \\
&= \;\; x^4 \;\;\;\; - \;\; 12x^3y \;\;\;\; + \;\; 54x^2y^2 \;\;\;\; - \;\; 108xy^3 \;\;\;\; + \;\; 81y^4
\end{aligned}
$$

KEY EXAMPLE

Write the expansion of $(x + y)^6$.

Solution: According to the binomial theorem:

$$(x + y)^6 = {}_6C_0\, x^6 + {}_6C_1\, x^5 y + {}_6C_2\, x^4 y^2 + {}_6C_3\, x^3 y^3 + {}_6C_4\, x^2 y^4 + {}_6C_5\, x y^5 + {}_6C_6\, y^6.$$

To evaluate the binomial coefficients using Pascal's triangle in the accompanying figure, form row 6 of the triangle by writing 1 as its first and last members. Find each of the other entries for row 6 by adding the two numbers directly above from row 5:

row 5 :		1	5	10	10	5	1	
row 6 :	1	6	15	20	15	6	1

Hence,

$$(x + y)^6 = 1x^6 + 6x^5 y + 15x^4 y^2 + 20 x^3 y^3 + 15 x^2 y^4 + 6 x y^5 + 1y^6.$$

Writing the *k*th term: Sometimes the only thing you need to know about the power of a binomial is what a particular term looks like. The *k*th term of the expansion of $(a + b)^n$ is given by the formula

$$k\text{th term of } (a + b)^n = {}_nC_{k-1}a^{n-(k-1)}b^{k-1}$$

where $k \leq n + 1$. For example, the expansion of $(x - 2y^3)^4$ contains $4 + 1 = 5$ terms. To find out what the middle or third term looks like, let $a = x$, $b = -2y$, $k = 3$, and $n = 4$. Thus,

$$\text{Third term of } (x - 2y^3)^4 = {}_4C_{3-1}(x)^{4-(3-1)}(-2y^3)^{3-1}$$
$$= {}_4C_2x^2(-2y^3)^2$$
$$= (6)(x^2)(4y^6)$$
$$= 24x^2y^6$$

KEY EXAMPLE

Find the numerical coefficient of the term that contains y^8x^5 in the expansion of $(y - 2x)^{13}$.

Solution: The first term contains y^{13}, the second term contains $y^{12}x$, the third term contains $y^{11}x^2$, and so forth. The sixth term contains y^8x^5.

$$\text{Sixth term of } (y - 2x)^{13} = {}_{13}C_{6-1}y^{13-(6-1)}(-2x)^{6-1}$$
$$= {}_{13}C_5y^8(-2x)^5$$
$$= (1287)y^8(-32)x^5$$
$$= -41184\,y^8x^5$$

Hence, the numerical coefficient of the sixth term is -41184.

Theme 11 SEQUENCES, SERIES, AND INDUCTIONS

*L*ists of numbers sometimes follow a recognizable pattern. Some number patterns that involve only positive integers can be generalized by statements that look like formulas. These formulas are proved using a special method called **mathematical induction**.

Key 64 Arithmetic sequences and series

OVERVIEW *A sequence such as 2, 5, 8, 11, 14, . . . is called an **arithmetic** sequence since each term after the first is obtained by adding a constant to the term that precedes it.*

Sequence (progression): A sequence is a list of numbers, called **terms**, written in a specific order. A **finite sequence** has a definite number of terms. An **infinite sequence** is nonending. An infinite sequence uses three trailing periods to indicate the pattern never ends, as in 2, 5, 8, 11, 14, ...

Common difference: In an arithmetic sequence, subtracting any term from the term that follows it always results in the same number. This constant difference is called the *common difference* and is denoted by d. Thus, 2, 5, 8, 11, 14, . . . is an arithmetic sequence in which $d = 3$ since $5 - 2 = 8 - 5 = 11 - 8 = \cdots = 3$.

Means: The terms of a sequence that are between any two nonconsecutive terms are the means. In the arithmetic sequence 2, 5, 8, 11, 14, . . . , the numbers 5, 8, and 11 are the arithmetic means between 2 and 14.

Series: The indicated sum of the terms of a sequence forms a series.

Sequence formulas: If a_1 denotes the first term of an arithmetic sequence with n terms, a_n denotes the last term, and S_n represents the sum of all the terms from a_1 to a_n, then

- $a_n = a_1 + (n - 1)d$ [Formula 1]

- $S_n = \dfrac{n}{2}(a_1 + a_n)$ [Formula 2]

- $S_n = \dfrac{n}{2}[2a_1 + (n - 1)d]$ [Formula 3]

KEY EXAMPLE

Insert three arithmetic means between −2 and 12.

Solution: You need to determine the three missing terms of an arithmetic sequence whose first term is −2 and whose last term is 12.

Substituting $n = 5$, $a_1 = -2$, and $a_5 = 12$ in formula 1 gives

$$12 = -2 + (5 - 1)d.$$

Solving this equation for d yields $d = 3.5$. Hence, the missing terms are 1.5, 5, and 8.5.

KEY EXAMPLE

If the eighth term of an arithmetic sequence is -18 and the third term is 7, find the sum of the first 30 terms.

Solution: There are six terms in the arithmetic subsequence in which $a_3 = 7$ is the first term and $a_8 = -18$ is the last term. By formula 1

$$-18 = 7 + (6 - 1)d.$$

Solving for d yields $d = -5$. Since $d = -5$ and $a_3 = 7$, $a_2 = 12$, and $a_1 = 17$. Now use formula 3 to find the sum:

$$S_{30} = \frac{30}{2}[2(17) + 29(-5)] = 1665.$$

KEY EXAMPLE

How many numbers from 15 to 633 are divisible by 3? What is the sum of these numbers?

Solution: The numbers from 15 to 633 that are divisible by 3 form the arithmetic sequence 15, 18, 21, . . . , 633. Thus, $a_1 = 15$, $a_n = 633$, and $d = 3$. By formula 1,

$$633 = 15 + (n - 1)3.$$

Solving for n gives $n = 207$. Hence, 207 numbers from 15 to 633 are divisible by 3. To find their sum, use formula 2, letting $a_1 = 15$, $a_n = 633$, and $n = 207$. Thus,

$$S_{207} = \frac{207}{2}(15 + 633) = 67,068.$$

Key 65 Geometric sequences and series

OVERVIEW   *A sequence such as 2, 8, 32, 128, . . . is called a **geometric** sequence since each term after the first is obtained by multiplying the term that precedes it by a nonzero constant different from 1.*

Common ratio: In a geometric sequence dividing any term after the first by the term that precedes it gives the same nonzero number. This constant quotient is called the *common ratio* and is denoted by r. Thus, 2, 8, 32, 128, . . . is a geometric sequence in which $r = 4$ since

$$\frac{8}{2} = \frac{32}{8} = \frac{128}{32} = \cdots = 4.$$

Formulas: If S_n represents the sum of all the terms of a geometric sequence between a_1 and a_n where $r \neq 1,$ then

- $a_n = a_1 r^{n-1}$ [Formula 4]

- $S_n = \dfrac{a_1(1 - r^n)}{1 - r}$ [Formula 5]

- $S_n = \dfrac{a_1 - a_n r}{1 - r}$ [Formula 6]

If the common ratio of an infinite geometric sequence is between -1 and 1, then the sum of the terms converges to a real number that can be found using the formula

- $S = \dfrac{a_1}{1 - r}$, where $|r| < 1$.   [Formula 7]

KEY EXAMPLE

Find the three geometric means between 36 and $\dfrac{64}{9}$, assuming $0 < r < 1$. Then find the sum of these five terms.

Solution: Since $0 < r < 1$, the terms of the geometric sequence are decreasing in value. Find r using formula 4. Substituting $n = 5$, $a_1 = 36$, and $a_5 = \dfrac{64}{9}$ into formula 4 yields

$$\frac{64}{9} = 36r^{5-1}.$$

Thus,

$$r^4 = \frac{16}{81}, \text{ so } r = \frac{2}{3}.$$

Hence, the three geometric means are $a_2 = \frac{2}{3}(36) = 24$, $a_3 = \frac{2}{3}(24) = 16$, and $a_4 = \frac{2}{3}(16) = \frac{32}{3}$. To find the sum of these five terms, use formula 6 letting $a_1 = 36$, $a_5 = \frac{64}{9}$, $r = \frac{2}{3}$, and $n = 5$. Thus,

$$S_5 = \frac{36 - \left(\frac{64}{9}\right)\left(\frac{2}{3}\right)}{1 - \frac{2}{3}} = \frac{36 - \frac{128}{27}}{\frac{1}{3}}$$

$$= \frac{27\left(36 - \frac{128}{27}\right)}{27\left(\frac{1}{3}\right)} = \frac{972 - 128}{9} = \frac{844}{9}$$

KEY EXAMPLE

Express the repeating decimal 0.131313. . . as the ratio of two integers.

Solution: Since 0.131313 . . . = 0.13 + 0.13(0.01) + 0.13(0.01)² + · · · , the repeating decimal 0.131313 . . . can be written as the sum of an infinite geometric sequence. Use formula 7, letting $a_1 = 0.13$ and $r = 0.01$. Thus,

$$0.131313\ldots = \frac{0.13}{1 - 0.01} = \frac{0.13}{0.99} = \frac{13}{99}.$$

Key 66 Generalized sequences

OVERVIEW *Rather than listing the first few terms of a sequence and hoping that the pattern is apparent, a sequence can be described using a formula-type expression.*

Sequence (function): A sequence is a function whose *domain* is a set of consecutive whole numbers that represent the position numbers of the terms in the sequence and whose *range* values are the actual **terms** of the sequence.

Notation: Instead of using the standard function notation $a(1)$, $a(2)$, $a(3)$, . . . , the terms or function values of a sequence may be represented using the subscripted variables $a_1, a_2, a_3, \ldots, a_n, \ldots$, where the position of a term is given by its subscript. The notation $\{a_n\}$ refers to the entire sequence in which a_n is the nth term.

> *Example:* In the sequence 10, 15, 20, and 25, the domain is the set of four position numbers, $\{1, 2, 3, 4\}$, and the range is $\{10, 15, 20, 25\}$. If a represents this sequence function, then $a_1 = 10$, $a_2 = 15$, $a_3 = 20$, and $a_4 = 25$.

n*th* Partial sum: The nth partial sum is the sum of the first n terms of a sequence, denoted by S_n. Thus,

$$S_n = a_1 + a_2 + a_3 + \cdots + a_n.$$

Explicit formulas: A sequence may be defined by an explicit formula that relates the general term a_n to its subscript n, as in $a_n = 2n + 1$. Thus, $a_1 = 2(1) + 1 = 3$, $a_2 = 2(2) + 1 = 5$, $a_3 = 2(3) + 1 = 7$, and so forth.

Recursion formula: A recursion formula is one that relates a general term of a sequence to one or more terms that precedes it. In the sequence 2, 4, 10, 28, . . . each term after the first is obtained by multiplying the preceding term by 3 and then subtracting 2 from the product. The *recursion* formula that expresses this number pattern is $a_n = 3(a_{n-1}) - 2$. For an arithmetic sequence whose common difference is d, the recursion formula is

$$a_n = a_{n-1} + d.$$

The recursion formula for a geometric sequence whose common ratio is r is

$$a_n = ra_{n-1}.$$

Sequences defined recursively: The recursion formula $a_n = 3(a_{n-1}) - 2$ does not by itself provide enough information to obtain each term of the sequence it defines. If we also know that $a_1 = 2$, we can obtain any term by direct substitution:

$$a_2 = 3(a_1) - 2 = 3(2) - 2 = 4$$

$$a_3 = 3(a_2) - 2 = 3(4) - 2 = 10$$

$$a_4 = 3(a_3) - 2 = 3(10) - 2 = 28$$

$$\vdots$$

A sequence is *defined recursively* when all of its terms can be computed from a recursion formula accompanied by the value(s) of one or more given terms.

Fibonacci sequence: A Fibonacci sequence is the sequence 1, 1, 2, 3, 5, 8, 13, . . . in which each term after the second is obtained by adding the two preceding terms. The sequence can be defined recursively by $a_n = a_{n-1} + a_{n-2}$ with $a_1 = a_2 = 1$.

Disadvantage of recursive formulas: Finding terms whose subscripts are large may require a great deal of tedious computation since each of the preceding terms also must be calculated.

Key 67   Summation notation

OVERVIEW   *The sum of any number of consecutive terms of a sequence can be indicated using the symbol Σ, which is the capital Greek letter sigma.*

Sigma notation: The notation $\displaystyle\sum_{i=1}^{n} a_i$ represents the sum of the numbers a_i as i takes on consecutive integer values from i equals 1 (the value at the bottom of sigma) to i equals n (the value at the top of sigma). Thus,

$$\sum_{i=1}^{n} a_i = a_1 + a_2 + a_3 + \cdots + a_n.$$

The sum $a_1 + a_2 + a_3 + \cdots + a_n$ is called the **expanded form** of $\displaystyle\sum_{i=1}^{n} a_i$ and variable i is called the **index variable** of this sum.

Example 1:   The arithmetic series $3 + 6 + 9 + 12 + 15 + 18$ can be represented as $\displaystyle\sum_{k=1}^{6} 3k$.

Example 2:   The geometric series $2 + 4 + 8 + 16 + 32$ can be represented as $\displaystyle\sum_{k=1}^{5} 2^k$.

Facts about summation notation:

- Although the index variable is always a nonnegative integer, it need not begin at 1. Thus,

$$\sum_{i=3}^{5} a_i = a_3 + a_4 + a_5.$$

- If a constant factor appears inside a summation sign, then it may be "passed through it." Thus,

$$\sum_{i=3}^{5} 2a_i = 2\sum_{i=3}^{5} a_i = 2(a_3 + a_4 + a_5).$$

- The terms that are being summed may be expressed in terms of the index variable. Thus,

$$\sum_{k=1}^{5} k^2 = 1^2 + 2^2 + 3^2 + 4^2 + 5^2 = 55.$$

- $\displaystyle\sum_{i=1}^{n} c = nc.$ For example, $\displaystyle\sum_{i=1}^{4} 3 = 4 \cdot 3 = 12$.

- $\displaystyle\sum_{i=1}^{n} (a_i \pm b_i) = \sum_{i=1}^{n} a_i \pm \sum_{i=1}^{n} b_i.$

KEY EXAMPLE

Represent each sum using sigma notation:

(a) $2^3 + 3^3 + 4^3 + \cdots + 10^3$

(b) $(-1) + (4) + (-9) + (16) + (-25) + \cdots + n^2$

Solution: (a) $\displaystyle\sum_{k=2}^{10} k^3 = 2^3 + 3^3 + 4^3 + \cdots + 10^3$

(b) $\displaystyle\sum_{k=1}^{n} (-1)^k k^2 = -1 + 4 - 9 + 16 - 25 + \cdots + n^2$

KEY EXAMPLE

Evaluate $\displaystyle\sum_{k=1}^{3} (2k - 1)^2.$

Solution: Add terms having the form $(2k - 1)^2$ for $k = 1, 2,$ and 3.

$$\sum_{k=1}^{3} (2k - 1)^2 = [(2 \cdot 1 - 1)^2 + (2 \cdot 2 - 1)^2 + (2 \cdot 3 - 1)^2]$$

$$= 1^2 + 3^2 + 5^2 = 35$$

KEY EXAMPLE

Evaluate $\displaystyle\sum_{k=1}^{\infty}\left(\frac{4}{7}\right)^{k}$.

Solution: The given sum represents an infinite geometric series whose first term is $\dfrac{4}{7}$ and whose common ratio is $\dfrac{4}{7}$. Since $\dfrac{4}{7} < 1$, formula 7 (page 186) can be used. Thus,

$$\sum_{k=1}^{\infty}\left(\frac{4}{7}\right)^{k} = \frac{\dfrac{4}{7}}{1-\dfrac{4}{7}} = \frac{4}{3}.$$

KEY EXAMPLE

Evaluate: (a) $\displaystyle\sum_{k=2}^{6}\left(\frac{3}{2}\right)^{k}$ (b) $\displaystyle\sum_{k=2}^{\infty}\left(\frac{3}{2}\right)^{k}$

Solution: (a) The given sum represents a geometric series with $n = 5$, $a_1 = \left(\dfrac{3}{2}\right)^{2} = \dfrac{9}{4}$, and $r = \dfrac{3}{2}$. Applying formula 5 (page 186) gives

$$\sum_{k=2}^{6}\left(\frac{3}{2}\right)^{k} = \frac{\left(\dfrac{9}{4}\right)\left(1-\left(\dfrac{3}{2}\right)^{5}\right)}{1-\dfrac{3}{2}} = \frac{\left(\dfrac{9}{4}\right)\left(1-\dfrac{243}{32}\right)}{-\dfrac{1}{2}} = \frac{\left(\dfrac{-1899}{128}\right)}{-\dfrac{1}{2}} = \frac{1899}{64}.$$

(b) Since the common ratio is greater than 1, the sum gets infinitely large and cannot be computed.

Key 68 Mathematical induction

OVERVIEW *Mathematical induction is a special method of mathematical proof that is particularly useful when it is necessary to prove formula-type statements that depend on a positive integer* n.

The need for induction: Let P_n represent a statement expressed in terms of a positive integer n, such as

$$P_n: 1 + 3 + 5 + 7 + \cdots + (2n - 1) = n^2,$$

where n is the number of terms in the series. You can easily verify that statements P_1, P_2, P_3, and P_4 are true:

P_1: $1 = 1^2$	P_3: $1 + 3 + 5 = 9 = 3^2$
P_2: $1 + 3 = 4 = 2^2$	P_4: $1 + 3 + 5 + 7 = 16 = 4^2$

Although statements P_1, P_2, P_3, and P_4 are true, can you be certain that P_n will continue to hold true for each successive integer replacement for n? Since it is not practicable or possible to test P_n for every possible integer value of n, a method for proving (or disproving) that this statement is always true is needed.

The domino principle of induction: Imagine an endless row of standing dominoes in which P_1 is written on the face of the first domino, P_2 is written on the face of the second domino, and so forth. All of the dominoes falling in succession, each pushing over the domino that follows, corresponds to the situation in which statements P_1, P_2, P_3, . . . , P_n are all true. All of the dominoes falling in succession requires that two conditions be met: (1) the first domino falls; and (2), if another domino after the first falls, it pushes over the domino that follows it.

Mathematical principle of induction: To prove by mathematical induction that the statement

$$P_n: 1 + 3 + 5 + 7 + \cdots + (2n - 1) = n^2$$

is true for all possible positive-integer replacements for n, complete these two steps:

- VERIFICATION STEP: Show that P_n is true when n is equal to its smallest possible value, which is called the **anchor value**. This step corresponds to verifying that the first domino falls. In this example, n can be any positive integer, so the anchor value is 1. When $n = 1$, statement P_n becomes $1 = 1^2$, which is true.

- INDUCTION STEP: Show that $P_{n=k+1}$ is true, assuming that $P_{n=k}$, where $k > 1$, is true. This step corresponds to showing that, when a domino after the first falls, the next domino also falls. The assumption that P_k is true is called the **induction hypothesis**. Assuming that P_k is true, you can write

P_k: $1 + 3 + 5 + 7 + \cdots + (2k - 1) = k^2$ ← Induction hypothesis.

Now show that P_{k+1} is also true. The next term on the left side of the equation would be $(2k - 1 + 2) = (2k + 1)$. Thus, adding $(2k + 1)$ to both sides of P_k produces an equivalent equation:

$1 + 3 + 5 + 7 + \cdots + (2k - 1) + \boxed{2k+1} = k^2 + \boxed{2k+1}$.

You can simplify the right side of the equation by factoring $k^2 + 2k + 1$ as $(k + 1)^2$:

$1 + 3 + 5 + 7 + \cdots + (2k - 1) + (2k + 1) = (k + 1)^2$.

The last equation is the statement P_{k+1} since it is the same statement obtained by replacing n with $k + 1$ in P_n. Because P_{k+1} is true when P_k is true, the induction step is finished.

Hence, by mathematical induction, P_n is true for all positive integers.

KEY EXAMPLE

Prove: $1 + 2 + 4 = 8 + \cdots + 2^{n-1} = 2^n - 1$ for all positive-integer values of n.

Solution: Complete the two steps required for an induction proof.

VERIFICATION STEP: The anchor value for n is 1. The statement P_1 is true because replacing n with 1 gives $1 = 2^1 - 1 = 1$, which is true.

INDUCTION STEP: Complete the induction step. Write the induction hypothesis by replacing n with k:

$$P_k: 1 + 2 + 4 + 8 + \cdots + 2^{k-1} = 2^k - 1.$$

Since the next consecutive term on the left side of the equation would be 2^k, add 2_k to both sides of the preceding equation:

$$1 + 2 + 4 + 8 + \cdots + 2^{k-1} + 2^k = 2^k + 2^k - 1.$$

On the right side of the equation, rewrite $2^k + 2^k$ as $2 \cdot 2^k = 2^1 \cdot 2^k = 2^{k+1}$. Then

$$1 + 2 + 4 + 8 + \cdots + 2^{k-1} + 2^k = 2^{k+1} - 1.$$

The last equation is the statement P_{k+1} since it is the same statement that is obtained by letting $n = k + 1P_n$.

Because both the verification and induction steps have been completed, the proof by mathematical induction is complete.

Proving divisibility: Mathematical induction is used in proofs other than those that establish general sum formulas. Some algebraic expressions that depend on a positive integer n are always divisible by the same constant. For example:

- $4^2 - 1$ is divisible by 3 since $\dfrac{4^2 - 1}{3} = \dfrac{15}{3} = 5$.

- $4^3 - 1$ is divisible by 3 since $\dfrac{4^3 - 1}{3} = \dfrac{63}{3} = 21$.

- $4^4 - 1$ is divisible by 3 since $\dfrac{4^4 - 1}{3} = \dfrac{255}{3} = 85$.

If P_n represents the statement "$4^n - 1$ is divisible by 3 for all positive-integer values of n," mathematical induction can be used to prove that statement P_n is always true. Here is the proof:

VERIFICATION STEP: The anchor value is 1. Statement P_1 is true because $4^1 - 1 = 3$ and 3 is divisible by 3.

INDUCTION STEP: Assume that the induction hypothesis, "$4^k - 1$ is divisible by 3," is true. If $4^k - 1$ is divisible by 3, then $4^k - 1$ must be a whole-number multiple of 3. Thus, $4^k - 1 = 3p$, where p is a whole number.

- Multiply both sides of $4^k - 1 = 3p$ by 4:

    $$4(4^k - 1) = 4(3p), \text{ so } 4^{k+1} - 4 = 12p.$$

- Add 3 to each side of $4^{k+1} - 4 = 12p$. The result is

    $$4^{k+1} - 1 = 12p + 3 \text{ or, equivalently, } 4^{k+1} - 1 = 3(4p + 1).$$

- Because $4p + 1$ is a positive integer, $3(4p + 1)$ is an integer multiple of 3, so $4^{k+1} - 1$ is divisible by 3. The last statement corresponds to the statement P_{k+1}.

Since P_{k+1} is true, the proof by mathematical induction is complete.

Proving a general inequality: Some pairs of algebraic expressions that depend on a positive-integer value of n maintain the same size relationship as n increases. For example, a comparison of $n!$ with 2^n for integer values of n greater than or equal to 4 is shown in the accompanying table.

$n!$	2^n	Size Relationship
$4! = 4 \times 3 \times 2 \times 1 = 24$	$2^4 = 16$	$4! > 2^4$
$5! = 5 \times 4 \times 3 \times 2 \times 1 = 120$	$2^5 = 32$	$5! > 2^5$
.	.	.
.	.	.
.	2^n	$n! > 2^n$
.		
$n!$		

The general inequality statement $n! > 2^n$ for $n \geq 4$ can be proved using mathematical induction:

VERIFICATION STEP: Since the anchor value is 4, you need to show that P_4 is true. Because $4! = 4 \times 3 \times 2 \times 1 = 24$ and $2^4 = 16$, $4! > 2^4$ is true.

INDUCTION STEP: Assume that the induction hypothesis, "$P_k : k! > 2^k$," is true.

* Multiply each side of $k! > 2^k$ by $(k + 1)$:

$$(k + 1)(k!) > (k + 1)(2^k).$$

Since $(k + 1)(k!) = (k + 1)!$, write the preceding inequality as $(k + 1)! > (k + 1)(2^k)$.

* Since the anchor value is 4, $k > 4$:

$$2^k = 2^k$$
$$\underline{\underline{(k + 1)}} \cdot 2^k > \underline{\underline{(2)}} \cdot 2^k \qquad \leftarrow \text{Because } k + 1 > 2$$
$$(k + 1) \cdot 2^k > 2^{k+1} \qquad \leftarrow \text{Rewrite } 2 \cdot 2k \text{ as } 2^{k+1}.$$

* You now know that $(k + 1)! > (k + 1)(2^k)$ and $(k + 1) \cdot 2^k > 2^{k+1}$. Hence, $(k + 1)! > 2^{k+1}$. The last inequality is the same inequality obtained by replacing n with $k + 1$ in P_n.

Hence, P_{k+1} is true, so the proof by mathematical induction is complete.

Theme 12 MATRICES AND DETERMINANTS

*S*ome advanced mathematical techniques depend on a rectangular arrangement of numbers, called a **matrix**, to help model situations that can be represented by a system of linear equations in two or more variables. Matrices provide an efficient way of organizing and manipulating large numbers of data values.

Key 69 Matrices

OVERVIEW *A rectangular arrangement of numbers that are arranged in* m *horizontal rows and* n *vertical columns forms an* m × n *(read as "*m *by* n*") **matrix**. The plural of* **matrix** *is "matrices."*

Matrix notation: A matrix is usually named by a boldfaced capital letter, and its terms, called **elements**, are enclosed by brackets.

Columns

$$
\begin{array}{cccc}
1 & 2 & 3 & 4 \\
\downarrow & \downarrow & \downarrow & \downarrow
\end{array}
$$

$$
\mathbf{A} = \begin{bmatrix} 2 & -5 & 6 & 9 \\ 8 & 1 & 7 & 3 \\ -3 & -4 & 0 & \sqrt{5} \end{bmatrix} \begin{array}{l} \rightarrow \ Row\ 1 \\ \rightarrow \ Row\ 2 \\ \rightarrow \ Row\ 3 \end{array}
$$

Matrix **A** is a 3 × 4 matrix since it has three rows and four columns. Thus, the **dimension** or size of this matrix is 3 × 4.

Locating an element of a matrix: To refer to an element of a matrix, give its row number followed by its column number. The element that occupies the ith row and jth column in matrix **A** is denoted by the double-subscripted variable a_{ij}. For instance, $a_{23} = 7$ and $a_{32} = -4$. Similarly, the element in row i and column j of **B** is denoted by b_{ij}. Sometimes the bracketed notation $[a_{ij}]$ is used to represent the entire array of elements of matrix **A**.

Square matrix: A square matrix has the same number of columns as rows. The **main diagonal** of a square matrix is the line that contains all elements whose row and column numbers are the same ($i = j$), as shown for the 3 × 3 matrix.

$$
\begin{bmatrix} a_{11} & a_{12} & a_{13} \\ a_{21} & a_{22} & a_{23} \\ a_{31} & a_{32} & a_{33} \end{bmatrix}
$$

Matrix addition and subtraction: Matrices that have the same dimension are added or subtracted by combining corresponding pairs of elements. Thus,

$$\begin{bmatrix} 5 & 7 \\ -2 & 4 \end{bmatrix} + \begin{bmatrix} 1 & -4 \\ 2 & 3 \end{bmatrix} = \begin{bmatrix} 5+1 & 7+(-4) \\ -2+2 & 4+3 \end{bmatrix} = \begin{bmatrix} 6 & 3 \\ 0 & 7 \end{bmatrix}.$$

The commutative and associative properties hold for matrix addition.

Scalar multiplication: To multiply a matrix by a number, multiply each element of the matrix by that number. Thus, $kA = k[a_{ij}] = [k\,a_{ij}]$ where k is some real number called the **scalar**. For instance,

$$2\begin{bmatrix} -1 & 3 \\ 4 & 0 \end{bmatrix} = \begin{bmatrix} 2(-1) & 2(3) \\ 2(4) & 2(0) \end{bmatrix} = \begin{bmatrix} -2 & 6 \\ 8 & 0 \end{bmatrix}.$$

Row by column multiplication: Multiplying a row of a matrix by a column of another matrix is accomplished by multiplying the first element in the row by the first element in the column, the second element in the row by the second element in the column, and so forth, and then adding these products. For instance,

$$\begin{bmatrix} 2 & 4 & -1 \end{bmatrix} \begin{bmatrix} 5 \\ 3 \\ 8 \end{bmatrix} = \begin{bmatrix} 2\cdot 5 + 4\cdot 3 + (-1)8 \end{bmatrix} = \begin{bmatrix} 14 \end{bmatrix}.$$

Matrix multiplication: Multiplying the ith row of matrix **A** by the jth column of matrix **B** gives the element in the ith row and jth column of their product matrix. For example,

$$\mathbf{AB} = \begin{bmatrix} 5 & -3 & 1 \\ -1 & 0 & 4 \end{bmatrix} \begin{bmatrix} 6 & 1 \\ -2 & 3 \\ 7 & -5 \end{bmatrix} = \begin{bmatrix} \text{row 1 of } \mathbf{A} \text{ times} & \text{row 1 of } \mathbf{A} \text{ times} \\ \text{column 1 of } \mathbf{B} & \text{column 2 of } \mathbf{B} \\ & \\ \text{row 2 of } \mathbf{A} \text{ times} & \text{row 2 of } \mathbf{A} \text{ times} \\ \text{column 1 of } \mathbf{B} & \text{column 2 of } \mathbf{B} \end{bmatrix}$$

$$= \begin{bmatrix} 5(6) + (-3)(-2) + 1(7) & 5(1) + (-3)(3) + 1(-5) \\ (-1)6 + 0(-2) + 4(7) & (-1)(1) + 0(3) + 4(-5) \end{bmatrix}$$

$$= \begin{bmatrix} 43 & -9 \\ 22 & -21 \end{bmatrix}$$

Compatibility of matrices: In order to be able to perform row by column matrix multiplication, the number of elements in each row of the first matrix must equal the number of elements in each column of second matrix. In general, matrices **A** and **B** are **compatible** for multiplication only if the number of columns of matrix **A** equals the number of rows of matrix **B**. If two matrices are not compatible, then they cannot be multiplied.

Dimension of a product matrix: If **A** is an $m \times n$ matrix and **B** is an $n \times p$ matrix, then **A** and **B** are compatible matrices and the product matrix **AB** is an $m \times p$ matrix. Thus, **AB** has the same number of rows as **A** and the same number of columns as **B**.

AB ≠ BA: The order in which two matrices are multiplied matters. The commutative property does *not* hold for matrix multiplication, although the associative and distributive properties do.

Identity matrix for multiplication: A square $n \times n$ matrix, denoted by $\mathbf{I}_n$, that has 1 in each position along its main diagonal and 0 in every other position is an identity matrix. Thus,

$$\mathbf{I}_2 = \begin{bmatrix} 1 & 0 \\ 0 & 1 \end{bmatrix} \quad \text{and} \quad \mathbf{I}_3 = \begin{bmatrix} 1 & 0 & 0 \\ 0 & 1 & 0 \\ 0 & 0 & 1 \end{bmatrix}.$$

The identity matrix plays the same role in matrix multiplication as 1 does in arithmetic multiplication. Thus, if **A** is a square matrix, then

$$\mathbf{A} \, \mathbf{I} = \mathbf{I} \, \mathbf{A} = \mathbf{A}.$$

The subscript n in $\mathbf{I}_n$ is omitted since we always assume it has the same size as the matrix it multiplies.

Key 70 Determinants

OVERVIEW *Every* n × n *square matrix* **A** *has an* n*th = order* ***determinant****, denoted by* |**A**|, *that associates a numerical value with the matrix.*

Second-order determinant: If **A** is the 2 × 2 square matrix $\begin{bmatrix} a_{11} & a_{12} \\ a_{21} & a_{22} \end{bmatrix}$, then |**A**| denotes a second-order determinant that is defined as the product of the elements along the main diagonal minus the product of the elements along the other diagonal:

$$|\mathbf{A}| = \begin{vmatrix} a_{11} & a_{12} \\ a_{21} & a_{22} \end{vmatrix} = (a_{11} \cdot a_{22}) - (a_{21} \cdot a_{12}).$$

For example,

$$\begin{vmatrix} 8 & 2 \\ 7 & 3 \end{vmatrix} = (8 \cdot 3) - (7 \cdot 2) = 24 - 14 = 10.$$

Minor of element a_{ij} of an nth-order determinant: The determinant of order $n - 1$, denoted by M_{ij}, formed by deleting the row and column that contains a_{ij}, is the minor of element a_{ij}. For the third-order determinant given at the right, the minor of 4 (= a_{12}) is the second-order determinant, denoted by M_{12}, obtained by crossing out the row and column that contains 4 in the original determinant. Thus,

$$\begin{vmatrix} 3 & 4 & -2 \\ 2 & -1 & -3 \\ 5 & 1 & 0 \end{vmatrix}$$

$$\begin{vmatrix} 3 & 4 & -2 \\ 2 & -1 & -3 \\ 5 & 1 & 0 \end{vmatrix} \longrightarrow \textit{Minor of 4} \longrightarrow \mathbf{M}_{12} = \begin{vmatrix} 2 & -3 \\ 5 & 0 \end{vmatrix} = 15.$$

Cofactor of element a_{ij} of an nth-order determinant: This value is denoted by $\mathbf{C}_{ij}$ where $\mathbf{C}_{ij} = (-1)^{i+j} \mathbf{M}_{ij}$. If

$$\mathbf{M}_{12} = 15,$$

then

$$\mathbf{C}_{ij} = (-1)^{1+2}(15) = -15.$$

To evaluate a third-order determinant: Write the determinant in terms of second-order determinants as follows:

$$\begin{vmatrix} a_{11} & a_{12} & a_{13} \\ a_{21} & a_{22} & a_{23} \\ a_{31} & a_{32} & a_{33} \end{vmatrix} = a_{11}\begin{vmatrix} a_{22} & a_{23} \\ a_{32} & a_{33} \end{vmatrix} - a_{21}\begin{vmatrix} a_{12} & a_{13} \\ a_{32} & a_{33} \end{vmatrix} + a_{31}\begin{vmatrix} a_{12} & a_{13} \\ a_{22} & a_{23} \end{vmatrix}.$$

Notice on the right side of the equation, that the expression that multiplies a_{11} is the cofactor of a_{11}, the expression that multiplies a_{21} (including the negative sign) is the cofactor of a_{21}, and the expression that multiplies a_{31} is the cofactor of a_{31}. Thus,

$$\begin{vmatrix} a_{11} & a_{12} & a_{13} \\ a_{21} & a_{22} & a_{23} \\ a_{31} & a_{32} & a_{33} \end{vmatrix} = a_{11}C_{11} + a_{21}C_{21} + a_{31}C_{33}.$$

The preceding equation gives the expansion of a third-order determinant along its *first* row since the right side of the equation represents the sum of the products of all elements in the first row and their cofactors. A determinant of any order may be expanded by cofactors along *any* row or column. If we choose to expand a third-order determinant A along its second column, then,

$$|\mathbf{A}| = a_{12}C_{12} + a_{22}C_{22} + a_{32}C_{32}.$$

KEY EXAMPLE

Evaluate: $|\mathbf{A}| = \begin{vmatrix} 3 & 4 & -2 \\ 2 & -1 & -3 \\ 5 & 1 & 0 \end{vmatrix}$

Solution: Evaluate the determinant by choosing any convenient row or column of the determinant and expanding by cofactors along this row or column. Expand along the third column since it contains 0. Thus,

$$\begin{aligned}
|\mathbf{A}| &= \quad (-2)C_{13} \quad + \quad (-3)C_{23} \quad + \; 0C_{33} \\
&= (-2)(-1)^4\begin{vmatrix} 2 & -1 \\ 5 & 1 \end{vmatrix} + (-3)(-1)^5\begin{vmatrix} 3 & 4 \\ 5 & 1 \end{vmatrix} + \quad 0 \\
&= \quad (-2)(7) \quad + \quad (3)(-17) \quad \\
&= \quad -14 + (-51) \\
&= \quad\quad -65
\end{aligned}$$

Determinants that evaluate to 0: A determinant will evaluate to 0 if it has any of the following properties:

- All the elements of any row or column are 0.
- Two rows or two columns are identical.
- A row (or column) is a multiple of another row (or column).

Operations that do not affect a determinant:
- Interchanging all the rows and columns.
- Replacing any row (or column) with the sum of that row (or column) and a fixed multiple of another row (or column).

Operations that affect a determinant:
- Interchanging any two rows or any two columns of a determinant changes the sign of the determinant.
- Multiplying each element in a row (or column) by some constant multiplies the value of the determinant by the same constant.

Cramer's rule: If the system of linear equations has exactly one solution, then the solution is given by

$$a_1 x + b_1 y = c_1$$
$$a_2 x + b_2 y = c_2$$

$$x = \frac{D_x}{D} \quad \text{and} \quad y = \frac{D_y}{D},$$

where

$$D = \begin{vmatrix} a_1 & b_1 \\ a_2 & b_2 \end{vmatrix} \neq 0, \quad D_x = \begin{vmatrix} c_1 & b_1 \\ c_2 & b_2 \end{vmatrix}, \quad \text{and} \quad D_y = \begin{vmatrix} a_1 & c_1 \\ a_2 & c_2 \end{vmatrix}.$$

Notice that D is the determinant whose members are the coefficients of the variables in the given system of equations, D_x is the determinant obtained from determinant D by replacing the coefficients of x with c_1 and c_2, and D_y is the determinant obtained from determinant D by replacing the coefficients of y with c_1 and c_2.

If $D = 0$, then Cramer's rule cannot be used. If $D = D_x = D_y = 0$, the system is dependent and has an infinite number of solutions. If $D = 0$, but at least one of D_x or D_y is *not* 0, then the system is inconsistent and there are no solutions. The following Key Example illustrates how Cramer's rule can be extended to solve a system of three linear equations.

KEY EXAMPLE

Use Cramer's rule to solve the following system of equations.

$$x - 2y + z = 3$$
$$2x - z = -7$$
$$y + 2z = 5$$

Solution: The determinant D of the coefficients is $D = \begin{vmatrix} 1 & -2 & 1 \\ 2 & 0 & -1 \\ 0 & 1 & 2 \end{vmatrix}$.

Replacing each column of this determinant with the constant terms gives

$$D_x = \begin{vmatrix} 3 & -2 & 1 \\ -7 & 0 & -1 \\ 5 & 1 & 2 \end{vmatrix}, \quad D_y = \begin{vmatrix} 1 & 3 & 1 \\ 2 & -7 & -1 \\ 0 & 5 & 2 \end{vmatrix}, \quad \text{and} \quad D_z = \begin{vmatrix} 1 & -2 & 3 \\ 2 & 0 & -7 \\ 0 & 1 & 5 \end{vmatrix}.$$

You can verify that $D = 11$, $D_x = -22$, $D_y = -11$, and $D_z = 33$. Thus,

$$x = \frac{D_x}{D} = \frac{-22}{11} = -2; \quad y = \frac{D_y}{D} = \frac{-11}{11} = -1; \quad \text{and} \quad z = \frac{D_z}{D} = \frac{33}{11} = 3.$$

Key 71 Triangularizing linear systems

OVERVIEW *A system of linear equations can be put into matrix form and then solved by arranging the elements of the matrix in a special triangular form in which all the elements below the main diagonal are 0.*

Triangular form of a three-equation linear system: A system of three equations in three unknowns, say x, y, and z, is in triangular form when variable z is the only variable in the last equation and variables y and z are the only variables in the middle equation.

Example: The system at the right is in triangular form.

$$\begin{aligned} 4x + 3y - z &= 1 \\ 2y + 3z &= 4 \\ -7z &= 14 \end{aligned}$$

Its solution can be obtained by solving the last equation for z so $z = -2$. Substituting this value of z in the middle equation and solving for y gives $y = 5$. Lastly, substituting the values for y and z in the first equation and solving for x gives $x = -4$. This method of solution is called "**back-substitution**."

Matrix form of a system of linear equations: To put a system of three equations in three unknowns in matrix form, write the first equation of the system in the form $Ax + By + Cz = k$. Then write the coefficients A, B, C, and the constant k for this equation as the elements of the first row of a matrix. Do the same for each of the other equations of the system. For example,

$$\begin{aligned} x + 2y + z &= 2 \\ 4x - y - 3z &= -4 \\ 2x + y - z &= -2 \end{aligned} \quad \text{corresponds to} \quad \left[\begin{array}{ccc|c} 1 & 2 & 1 & 2 \\ 4 & -1 & -3 & -4 \\ 2 & +1 & -1 & -2 \end{array}\right].$$

The 3×3 matrix to the left of the vertical dashed line is the **coefficient matrix.** The 3×4 matrix is called the **augmented matrix** since it includes the constant terms as well as the numerical coefficients of the variables. Each row of the augmented matrix corresponds to an equation of the system. If an equation of the system does not contain a particular variable, then its coefficient is 0 which is used as the corresponding element in the augmented matrix.

Row equivalent matrices: Row equivalent matrices are obtained from a given matrix by using one or more of the following **elementary row operations**:

1. Interchanging two rows.
2. Multiplying or dividing a row by a nonzero constant.
3. Replacing a row with the sum of that row and a multiple of another row.

A row equivalent matrix corresponds to a new system of linear equations that is equivalent to the system of equations that the original matrix represents.

Matrix solution of a linear system of equations: Form the augmented matrix that represents the given system of linear equations. Convert this matrix into a row equivalent matrix in triangular form. Then translate each row of the triangular matrix into a linear equation and solve the resulting system of equations by back-substitution.

KEY EXAMPLE

Solve the following system of equations:

$$\begin{aligned}
x + 2y + z &= 2 \\
4x - y - 3z &= -4 \\
2x + y - z &= -2
\end{aligned}$$

Solution: Form the augmented matrix and then convert it into a row equivalent matrix in *triangular form*. Replace the second row (R2) with the sum of that row (R2) and the third row (R3) multiplied by –2. Thus,

$$\left[\begin{array}{ccc|c}
1 & 2 & 1 & 2 \\
4 & -1 & -3 & -4 \\
2 & +1 & -1 & -2
\end{array}\right]
\quad R2 \to +R2 + (-2)R3 \quad
\left[\begin{array}{ccc|c}
1 & 2 & 1 & 2 \\
0 & -3 & -1 & 0 \\
2 & +1 & -1 & -2
\end{array}\right]$$

Replace the third row (R3) by the sum of that row (R3) and the first row (R1) multiplied by –2 .

$$\left[\begin{array}{ccc|c}
1 & 2 & 1 & 2 \\
0 & -3 & -1 & 0 \\
2 & +1 & -1 & -2
\end{array}\right]
\quad R3 \to R3 + (-2)R1 \quad
\left[\begin{array}{ccc|c}
1 & 2 & 1 & 2 \\
0 & -3 & -1 & 0 \\
0 & -3 & -3 & -6
\end{array}\right]$$

Replace the third row by the sum of that row and the second row multiplied by –1. The matrix that results will be in triangular form:

$$\begin{bmatrix} 1 & 2 & 1 & \vdots & 2 \\ 0 & -3 & -1 & \vdots & 0 \\ 0 & -3 & -3 & \vdots & -6 \end{bmatrix} \quad R3 \rightarrow R3 + (-1)R2 \quad \begin{bmatrix} 1 & 2 & 1 & \vdots & 2 \\ 0 & -3 & -1 & \vdots & 0 \\ 0 & 0 & -2 & \vdots & -6 \end{bmatrix}$$

Once the matrix is in triangular form, it can be converted back into an equivalent system of equations that can be solved by back-substitution. Solving the last equation gives $z = 3$.

$$\begin{aligned} x + 2y + z &= 2 \\ -3y - z &= 0 \\ -2z &= -6 \end{aligned}$$

Substituting into the second equation and solving for y gives $y = -1$. Substituting the solutions for y and z in the first equation and solving for x gives $x = 1$.

A linear system with infinitely many solutions: After triangularizing a linear system, one of its equations may have the form of 0 times a variable equals 0. We can arbitrarily assign this variable any value and then calculate the corresponding values of other variables.

Example: In the triangularized system of equations shown at the right, the last equation tells us that z can have *any* value. Solving the second equation for y gives $y = 5 - z$.

$$\begin{aligned} x + y - 2z &= 9 \\ y + z &= 5 \\ 0z &= 0 \end{aligned}$$

Substituting this expression for y in the first equation and solving for x gives $3z + 4$. Hence, the infinite set of solutions for the original system of equations can be represented as

$$x = 3z + 4, \quad y = 5 - z, \quad \text{and} \quad z = \text{any real number.}$$

For example, if $z = 1$, then $x = 3(1) + 4 = 7$, and $y = 5 - 1 = 4$.

A linear system with no solutions: Consider the system of three equations from the preceding example, but with the equation $0z = 0$ replaced by $y + z = -3$. The last equation now contradicts the second equation, so the system has no solutions.

Key 72   Finding the inverse of a matrix

OVERVIEW   *Since* $5 \times \dfrac{1}{5} = 1$, $\dfrac{1}{5}$ *is called the multiplicative inverse of 5. Similarly, if* $AB = I$, *then* B *is called the inverse of matrix* A. *The inverse of matrix* A *is usually written as* A^{-1}. *A matrix may or may not have an inverse.*

Inverse of a square matrix A: The inverse of matrix **A** is the square matrix denoted by A^{-1}, provided that it exists, that has the property

$$A\,A^{-1} = A^{-1}A = I.$$

Nonsingular matrix: A square matrix that has an inverse matrix is nonsingular.

Finding the inverse of a matrix: If **A** is an $n \times n$ nonsingular matrix, then its inverse is obtained by following these steps:

1. Append to matrix **A** the identity matrix I_n so that the resulting $n \times 2n$ matrix has the form $[A \vdots I_n]$.
2. Use elementary row operations to transform this matrix into a row equivalent matrix with the form $[I_n \vdots B]$.
3. Form matrix **B**, which is the inverse of **A**; that is, $B = A^{-1}$.

KEY EXAMPLE

Find the inverse of matrix **A** where $A = \begin{bmatrix} 1 & 2 \\ 3 & -1 \end{bmatrix}$.

Solution: Append the 2×2 identity matrix to matrix **A**:

$$\begin{bmatrix} 1 & 2 & \vdots & 1 & 0 \\ 3 & -1 & \vdots & 0 & 1 \end{bmatrix}$$

Replace the second row (R2) by the sum of that row (R2) and the first row (R1) multiplied by –3. Then replace the second row (R2) by that row multiplied by $-\dfrac{1}{7}$. Thus,

$$\begin{bmatrix} 1 & 2 & \vdots & 1 & 0 \\ 3 & -1 & \vdots & 0 & 1 \end{bmatrix} \xrightarrow[\;R2 \to \left(-\frac{1}{7}\right)R2\;]{R2 \to R2 + (-3)R1} \begin{bmatrix} 1 & 2 & \vdots & 1 & 0 \\ 0 & 1 & \vdots & \dfrac{3}{7} & \dfrac{-1}{7} \end{bmatrix}$$

Replace the first row (R1) by the sum of that row (R1) and the second row (R2) multiplied by –2. Thus,

$$\begin{bmatrix} 1 & 2 & \vdots & 1 & 0 \\ 0 & 1 & \vdots & \dfrac{3}{7} & \dfrac{-1}{7} \end{bmatrix} \xrightarrow{R1 \to R1 + (-2)R2} \begin{bmatrix} 1 & 0 & \vdots & \dfrac{1}{7} & \dfrac{2}{7} \\ 0 & 1 & \vdots & \dfrac{3}{7} & \dfrac{-1}{7} \end{bmatrix}$$

Thus, $\mathbf{A}^{-1} = \begin{bmatrix} \dfrac{1}{7} & \dfrac{2}{7} \\ \dfrac{3}{7} & \dfrac{-1}{7} \end{bmatrix}$.

Key 73 Solving linear systems

by matrix inversion

OVERVIEW   *In algebra, if* ax = b, *then multiplying each side of the equation by the multiplicative inverse (reciprocal) of a gives the solution* $x = \dfrac{b}{a} = a^{-1} b$. *Similarly, if* **AX = K**, *then* **X = A⁻¹K**.

Representing a linear system as a matrix product: A system of linear equations can be represented by a *single* matrix equation. A linear system of equations having the form may be written as the matrix equation **AX = K**, where

$$a_1 x + b_1 y + c_1 z = k_1$$
$$a_2 x + b_2 y + c_2 z = k_2$$
$$a_3 x + b_3 y + c_3 z = k_3,$$

$$\mathbf{A} = \begin{bmatrix} a_1 & b_1 & c_1 \\ a_2 & b_2 & c_2 \\ a_3 & b_3 & c_3 \end{bmatrix}, \quad \mathbf{X} = \begin{bmatrix} x \\ y \\ z \end{bmatrix}, \quad \text{and} \quad \mathbf{K} = \begin{bmatrix} k_1 \\ k_2 \\ k_3 \end{bmatrix}.$$

Thus, the matrix equation **X = A⁻¹K** represents the solution to the original system of linear equations. The solution for any system of n linear equations in n unknowns, provided it exists, can be represented in this way. The following Key Example illustrates how to obtain the solution to a two-equation linear system using matrix inversion.

KEY EXAMPLE

Solve the system

$$x + 2y = -3$$
$$3x - y = 5$$

Solution: The given system of equations can be represented by the matrix equation

$$AX = K,$$

where

$$A = \begin{bmatrix} 1 & 2 \\ 3 & -1 \end{bmatrix}, \quad X = \begin{bmatrix} x \\ y \end{bmatrix}, \quad \text{and} \quad K = \begin{bmatrix} -3 \\ 5 \end{bmatrix}.$$

From the example in Key 72 (page 208),

$$A^{-1} = \begin{bmatrix} \dfrac{1}{7} & \dfrac{2}{7} \\ \dfrac{3}{7} & \dfrac{-1}{7} \end{bmatrix}.$$

Since $X = A^{-1}K$,

$$X = \begin{bmatrix} \dfrac{1}{7} & \dfrac{2}{7} \\ \dfrac{3}{7} & \dfrac{-1}{7} \end{bmatrix} \begin{bmatrix} -3 \\ 5 \end{bmatrix} = \begin{bmatrix} \dfrac{-3}{7} + \dfrac{10}{7} \\ \dfrac{-9}{7} + \dfrac{-5}{7} \end{bmatrix} = \begin{bmatrix} 1 \\ -2 \end{bmatrix} = \begin{bmatrix} x \\ y \end{bmatrix}.$$

Hence, $x = 1$ and $y = -2$.

Appendix
KEY TRIGONOMETRIC FACTS

Key Fact 1: An **angle** is a measure of rotation, with the measure of one complete rotation about a fixed point defined as 360°.

Key Fact 2: One **radian** is the measure of the angle at the center of a circle whose sides cut off on the circle an arc that has the same length as the radius r of the circle, as shown in Figure 1. Unlike degrees, radians are real numbers.

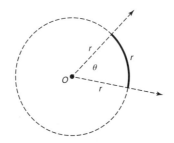

Figure 1 Angle θ, Where θ Is the
Greek Letter Theta, Measures 1 Radian

Key Fact 3: To convert from degrees to radians, multiply the number of degrees by $\frac{\pi}{180°}$:

$$60° = \overset{3}{\cancel{60°}} \times \frac{\pi}{\cancel{180°}} = \frac{\pi}{3} \text{ radians.}$$

Key Fact 4: To convert from radians to degrees, multiply the number of radians by $\frac{180°}{\pi}$:

$$\frac{7}{12}\pi \text{ radians} = \frac{7\cancel{\pi}}{\cancel{12}} \times \frac{\overset{15°}{\cancel{180°}}}{\cancel{\pi}} = 105°.$$

Key Fact 5: A trigonometric function relates the measures of *two* sides and one of the acute angles of a right triangle, as shown in Figure 2 and Table 1.

Table 1 The Six Trigonometric functions of Acute $\angle A$

Three Basic Trigonometric Functions: Sine, Cosine, and Tangent	Reciprocal Trigonometric Functions: Cosecant, Secant, and Cotangent
$\sin A = \dfrac{\text{leg opposite } \angle A}{\text{hypotenuse}} = \dfrac{a}{c}$	$\csc A = \dfrac{c}{a}$, so $\csc A = \dfrac{1}{\sin A}$
$\cos A = \dfrac{\text{leg adjacent to } \angle A}{\text{hypotenuse}} = \dfrac{b}{c}$	$\sec A = \dfrac{c}{b}$, so $\sec A = \dfrac{1}{\cos A}$
$\tan A = \dfrac{\text{leg opposite } \angle A}{\text{leg adjacent to } \angle A} = \dfrac{a}{b}$	$\cot A = \dfrac{b}{a}$, so $\cot A = \dfrac{1}{\tan A}$

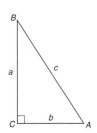

Figure 2 Right Triangle Trigonometry

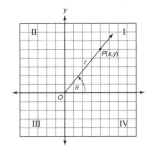

Figure 3 Defining an Angle Using Coordinates

Key Fact 6: If an angle is placed in the coordinate plane so that its vertex is at the origin and one side is fixed on the horizontal axis, as shown in Figure 3, the six trigonometric functions can be defined in terms of the coordinates of any point $P(x,y)$ on the side of the angle that rotates. The coordinate definitions are given in Table 2 where r is the distance of P from the origin, and a positive angle corresponds to a counterclockwise rotation of $\overrightarrow{OP}$, called the **terminal side** of the angle.

Table 2 Coordinate Definitions of the Six Trigonometric Functions

Coordinate Definitions ($x^2 + y^2 = r^2$)	Functions Are Positive Only in:
$\sin \theta = \dfrac{y}{r}$ and $\csc \theta = \dfrac{r}{y}$	Quadrants I and II
$\cos \theta = \dfrac{x}{r}$ and $\sec \theta = \dfrac{r}{x}$	Quadrants I and IV
$\tan \theta = \dfrac{y}{x}$ and $\cot \theta = \dfrac{x}{y}$	Quadrants I and III

Key Fact 7: For each angle θ in standard position, there is a corresponding acute angle (between 0 and 90°) called the *reference angle*. The **reference angle** θ_R is the *acute* angle whose vertex is the origin and whose sides are the terminal side of angle θ and the *x*-axis. If angle θ is a Quadrant I angle, then θ and θ_R are the same angle. Figure 4 shows how to locate the reference angle when angle θ terminates in Quadrant II, III, or IV. The right triangle that contains the reference angle is the **reference triangle**.

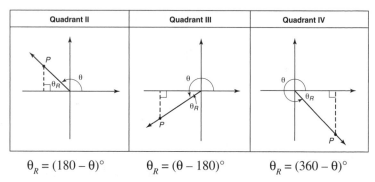

Quadrant II	Quadrant III	Quadrant IV
$\theta_R = (180 - \theta)°$	$\theta_R = (\theta - 180)°$	$\theta_R = (360 - \theta)°$

Figure 4 Determining Reference Angles

Key Fact 8: If a trigonometric function *f* of an angle θ is *positive* in the quadrant in which θ terminates, then $f(\theta) = f(\theta_R)$. Because sine is positive in Quadrant II, sin 135° = sin(180 − 135)° = sin 45°. See Figure 5. If a trigonometric function *f* is *negative* in the quadrant in which θ terminates, then $f(\theta) = -f(\theta_R)$. Thus, cos 242° = −cos(242 − 180)° = −cos 62°.

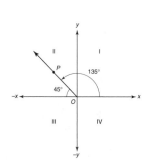

Figure 5. Evaluating a Function of an Angle Greater than 90°

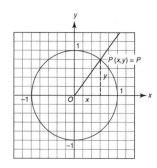

Figure 6. The Unit Circle

Key Fact 9: The **unit circle** is the circle centered at the origin whose radius is 1, as shown in Figure 6. If the terminal side of angle θ intersects the unit circle at $P(x,y)$, then

$$\cos\theta = \frac{x}{1} = x \quad \text{and} \quad \sin\theta = \frac{y}{1} = y, \text{ so } P(x,y) = P(\cos\theta, \sin\theta).$$

Key Fact 10: In the unit circle $\sin\theta = y$, $\cos\theta = x$, and $x^2 + y^2 = 1$. Since $\dfrac{\sin\theta}{\cos\theta} = \dfrac{y}{x}$ and $\tan\theta = \dfrac{y}{x}$, it follows that

$$\tan\theta = \frac{\sin\theta}{\cos\theta} \quad \text{and} \quad \cot\theta = \frac{\cos\theta}{\sin\theta}$$

for all values of θ that do not make the denominator evaluate to 0.

Key Fact 11: If $P(\cos\theta, \sin\theta)$ represents any point on the unit circle, then substituting $\cos\theta$ for x and $\sin\theta$ for y in $x^2 + y^2 = 1$ leads to the Pythagorean identities summarized in Table 3.

Table 3 Pythagorean Trigonometric Identities

Pythagorean Identities	Some Equivalent Forms
• $\sin^2\theta + \cos^2\theta = 1$ • $\tan^2\theta + 1 = \sec^2\theta$ • $\cot^2\theta + 1 = \csc^2\theta$	• $\sin^2\theta = 1 - \cos^2\theta$ *or* $\cos^2\theta = 1 - \sin^2\theta$ • $\tan^2\theta = \sec^2\theta - 1$ *or* $\sec^2\theta - \tan^2\theta = 1$ • $\cot^2\theta = \csc^2\theta - 1$ *or* $\csc^2\theta - \cot^2\theta = 1$

Key Fact 12: The trigonometric functions of $30°$, $45°$, and $60°$ have exact values, as shown in Table 4. To find the values of cosecant, secant, and cotangent of $30°$, $45°$, and $60°$, use the reciprocal relationships.

Table 4 Trigonometric Function Values of $30°$, $45°$, and $60°$

x	$\sin x$	$\cos x$	$\tan x$
$30°$	$\dfrac{1}{2}$	$\dfrac{\sqrt{3}}{2}$	$\dfrac{1}{\sqrt{3}}$ or $\dfrac{\sqrt{3}}{3}$
$45°$	$\dfrac{\sqrt{2}}{2}$	$\dfrac{\sqrt{2}}{2}$	1
$60°$	$\dfrac{\sqrt{3}}{2}$	$\dfrac{1}{2}$	$\sqrt{3}$

Key Fact 13: Angles greater than 90° or less than 0° may have reference angles of 30°, 45°, or 60°. For example:

- $\cos 120° = -\cos 60° = -\dfrac{1}{2}$

- $\sin 300° = -\sin 60° = -\dfrac{\sqrt{3}}{2}$

- $\tan 225° = \tan 45° = 1$

- $\cos(-30)° = \cos 30° = \dfrac{\sqrt{3}}{2}$

- $\csc 240° = -\dfrac{1}{\sin 60°} = -\dfrac{2}{\sqrt{3}}$

- $\sec 405° = \dfrac{1}{\cos 45°} = \sqrt{2}$

Key Fact 14: A **quadrantal angle** is an angle whose terminal side coincides with a coordinate axis. The values of sine, cosine, and tangent of the quadrantal angles are summarized in Table 5.

Table 5 Evaluating Trigonometric Functions of Quadrantal Angles

Trigonometric Function	0° or 360° = 2π	90° = $\dfrac{\pi}{2}$	180° = π	270° = $\dfrac{3}{2}\pi$
$\sin x$	0	1	0	−1
$\cos x$	1	0	−1	0
$\tan x$	0	Undefined	0	Undefined

To find the values of cosecant, secant, and cotangent of the quadrantal angles, use either the reciprocal or quotient function relationships. For example:

- $\sec 0° = \dfrac{1}{\cos 0°} = \dfrac{1}{1} = 1.$

- $\csc \pi = \dfrac{1}{\sin \pi} = \dfrac{1}{0}$, so $\csc \pi$ is undefined.

- $\cot 90° = \dfrac{\cos 90°}{\sin 90°} = \dfrac{0}{1} = 0.$

Key Fact 15: To solve a trigonometric equation, first solve for the trigonometric function. Then find the angle, keeping in mind that a trigonometric function is positive or negative in more than one quadrant. For example, if $\tan^2 x = \tan x + 2$, then $\tan^2 x - \tan x - 2 = 0$, so $(\tan x + 1)(\tan x - 2) = 0$ and $\tan x = -1$ or $\tan x = 2$.

- If $\tan x = -1$, the reference angle is $45°$. Tangent is negative in Quadrants II and III. Thus:

$$Q_{II}: x_1 = 180° - 45° = 135° \quad \text{or} \quad Q_{IV}: x_2 = 360° - 45° = 315°.$$

- If $\tan x = 2$, x is the angle whose tangent is 2, which may be written, using inverse trigonometric notation, as $x = \tan^{-1} 2$. Use the inverse tangent function of a calculator to find that the reference angle, to the *nearest tenth* of a degree, is $63.4°$. Because tangent is positive in Quadrants I and III:

$$Q_I: x_3 \approx 63.4° \quad \text{or} \quad Q_{III}: x_4 \approx 180° + 63.4° = 243.4°.$$

Key Fact 16: The area of a triangle is one-half of the product of the lengths of two sides and the sine of the included angle. Thus, in $\triangle ABC$,

$$\text{Area} = \frac{1}{2} \times b \times c \sin A = \frac{1}{2} \times a \times c \sin B = \frac{1}{2} \times a \times b \sin C.$$

Key Fact 17: The **Law of Sines** states that in a triangle the ratio of the length of any side of the triangle to the sine of the angle opposite that side is constant. Thus, in $\triangle ABC$,

$$\frac{a}{\sin A} = \frac{b}{\sin B} = \frac{c}{\sin C}.$$

The Law of Sines is needed to find the measure of a side or angle of a triangle, given **Angle-Angle-Side** (*AAS*), **Side-Side-Angle** (*SSA*), or **Angle-Side-Angle** (*ASA*) triangle measurements.

Key Fact 18: The Law of Cosines relates the cosine of any angle of a triangle to the lengths of the three sides of the triangle, as shown in Table 6. The Law of Cosines is used when **Side-Angle-Side** (*SAS*) or **Side-Side-Side** (*SSS*) triangle measurements are given.

Table 6 Law of Cosines, Given SAS

$\triangle ABC$	Given *SAS*	To Find	Law of Cosines
	$b, \angle A, c$	a	$a^2 = b^2 + c^2 - 2bc \cos A$
	$a, \angle B, c$	b	$b^2 = a^2 + c^2 - 2ac \cos B$
	$a, \angle C, b$	c	$c^2 = a^2 + b^2 - 2ab \cos C$

Key Fact 19: The sine, cosine, and tangent of the sum or difference of two angles can be written in terms of trigonometric functions of the individual angles, as summarized in Table 7.

Table 7 Sum and Difference Formulas

Sines, Cosines, and Tangents
• $\sin(A + B) = \sin A \cos B + \cos A \sin B$ $\sin(A - B) = \sin A \cos B - \cos A \sin B$
• $\cos(A + B) = \cos A \cos B - \sin A \sin B$ $\cos(A - B) = \cos A \cos B + \sin A \sin B$
• $\tan(A + B) = \dfrac{\tan A + \tan B}{1 - \tan A \cdot \tan B}$ $\tan(A - B) = \dfrac{\tan A - \tan B}{1 + \tan A \cdot \tan B}$

Key Fact 20: $\sin(-\theta) = -\sin\theta$, $\tan(-\theta) = -\tan\theta$, and $\cos(-\theta) = \cos\theta$.

Key Fact 21: The double-angle formulas $\sin 2A$, $\cos 2A$, and $\tan 2A$ can be derived from the corresponding formulas for $\sin(A + B)$, $\cos(A + B)$, and $\tan(A + B)$ by letting $B = A$. The results are given in Table 8.

Table 8 Double-Angle Formulas

• $\sin 2A = 2\sin A \cos A$

• $\cos 2A = \cos^2 A - \sin^2 A$
  or
• $\cos 2A = 2\cos^2 A - 1$
  or
• $\cos 2A = 1 - 2\sin^2 A$

• $\tan 2A = \dfrac{2\tan A}{1 - \tan^2 A}$

Key Fact 22: The half-angle formulas for sine, cosine, and tangent can be developed from the double-angle formulas for cosine and tangent by letting $2A = x$, so $A = \dfrac{1}{2}x$. See Table 9.

Table 9 Half-Angle Formulas

$$\sin \frac{1}{2}x = \pm\sqrt{\frac{1 - \cos x}{2}}$$

$$\cos \frac{1}{2}x = \pm\sqrt{\frac{1 + \cos x}{2}}$$

$$\tan \frac{1}{2}x = \pm\sqrt{\frac{1 - \cos x}{1 + \cos x}}$$

The choice of a positive or negative sign in front of each radical depends on the sign of the trigonometric function in the quadrant in which $\frac{1}{2}x$ lies. For example, if $270° \le x < 360°$, then $135° \le \frac{1}{2}x < 180°$, so $\frac{1}{2}x$ lies in Quadrant II. Thus, $\sin \frac{1}{2}x$ takes a positive sign, while $\cos \frac{1}{2}x$ and $\tan \frac{1}{2}x$ are both negative.

Key Fact 23: Trigonometric functions are **periodic** because there is a repeating pattern of the function values that is easily observed from their graphs, as shown in Figures 7 and 8. The length of the smallest interval of x-values that the function needs to complete one cycle is called the *period* of the function. Sine and cosine functions have periods of 2π, while the period of the tangent function is π.

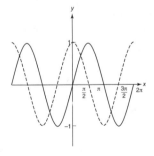

Figure 7 Graphs of $y = \sin x$ and $y = \cos x$

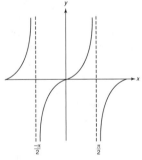

Figure 8 Graph of $y = \tan x$

GLOSSARY

Included here are the definitions of many, but not all, of the terms used in the Keys. For terms not listed here, please consult the index for page references.

abscissa   The x-coordinate of a point in the coordinate plane.

absolute value   The absolute value of a number x, denoted by $|x|$, is its distance from zero on the number line. If x is nonnegative, $|x| = x$ and, if x is negative, $|x| = -x$.

additive identity   In the set of real numbers, 0 is the additive identity element since the sum of any real number and 0 is that number.

additive inverse   In the set of real numbers, the additive inverse of any real number x is its opposite, $-x$, since $x + (-x) = 0$.

antilogarithm   The number whose logarithm is given. For example, if log 3 = 0.4771, then 3 is the antilogarithm of 0.4771. In general, if $\log_b x = y$, then x is the antilogarithm of y to the base b.

arithmetic sequence   A sequence of numbers or terms in which each term after the first is obtained from the term that precedes it by adding the same constant, called the *common difference*. The nth term of an arithmetic sequence is

$$a_n = a_1 + (n-1)d$$

where a_1 is the first term and d is the common difference.

arithmetic series   The indicated sum of the terms of an arithmetic sequence. The sum of the first n terms of an arithmetic sequence is

$$S_n = \frac{n}{2}(a_1 + a_n).$$

associative law   The order in which three real numbers are added or multiplied does not matter.

asymptote   A line that a graph approaches, but does not intersect, as x increases or decreases without bound.

axis of symmetry   A line that divides a graph so that if the graph is folded on the line the two parts of the graph will exactly coincide.

binomial   A polynomial with two unlike terms.

binomial theorem   A formula that tells how to expand a binomial of the form $(a + b)^n$ without performing repeated multiplications.

change of base formula   The formula $\log_a c = \dfrac{\log_b c}{\log_b a}$ changes the base of $\log_a c$ from a to b.

characteristic   The integer part of a common logarithm.

circle   The set of points (x, y) in a plane that are the same distance r from a given point (h, k) called the *center*. Thus, $(x - h)^2 + (y - k)^2 = r^2$.

closure property   A set of numbers is closed under an operation if the number that the operation produces is a member of the same set.

coefficient   The numerical factor of a monomial.

combination   A subset of a set of objects in which order does not matter.

combination formula   The combination of n objects taken r at a time, denoted by $_nC_r$, is given by the formula

$$_nC_r = \frac{_nP_r}{r!} = \frac{n!}{r!(n-r)!}.$$

common difference   In an arithmetic sequence, the constant that is added to each term after the first in order to obtain the next consecutive term.

common logarithm   A logarithm whose base is 10.

common ratio   In a geometric sequence, the constant that multiplies each term after the first in order to obtain the next consecutive term.

commutative law   The order in which two real numbers are added or multiplied does not matter.

complex fraction   A fraction that contains other fractions in its numerator, denominator, or in both the numerator and denominator.

complex number   A number that can be written in the form $a + bi$ where a and b are real numbers and $i = \sqrt{-1}$.

composite function   The composite function $f \circ g$ is the function that consists of the set of function values $f(g(x))$, provided $g(x)$ is in the domain of function f.

conic section   A curve formed when a plane intersects a right circular cone. Circles, ellipses, hyperbolas, and parabolas are conic sections.

conjugate axis   The axis of symmetry of a hyperbola that is perpendicular to the transverse axis at its center. The length of the conjugate axis is denoted by $2b$.

conjugate pair   The sum and difference of the same two terms, as in $a + \sqrt{b}$ and $a - \sqrt{b}$ or $a + bi$ and $a - bi$.

consistent system   A system of linear equations that has a unique solution.

constant   A quantity that is fixed in value. In the equation $y = 3x$, x and y are variables and 3 is a constant.

coordinate plane   The region formed when a horizontal number line and a vertical number line intersect at their zero points.

counting principle   If activity A can occur in m ways and activity B can occur in n ways, then both events can occur in m times n ways.

Cramer's rule   A method that uses determinants to solve a system of linear equations.

degree of a monomial   The sum of the exponents of the variable factors of the monomial.

degree of a polynomial   The greatest degree of the monomial terms of the polynomial.

determinant of a 2 × 2 matrix   If matrix $\mathbf{A} = \begin{bmatrix} a & c \\ b & d \end{bmatrix}$, then the determinant of $\mathbf{A}$, denoted by $|\mathbf{A}|$, is $a \cdot d - b \cdot c$.

dependent variable   For a function of the form $y = f(x)$, y is the dependent variable.

dimension   The size of a matrix determined by the number of its rows and columns.

direct variation   Variable y varies directly as x if $\dfrac{y}{x} = k$ (or $y = kx$) where k is called the *constant of variation*.

discriminant   For a quadratic equation $ax^2 + bx + c = 0$, the quantity $b^2 - 4ac$.

distance formula   The distance between the points (x_1, y_1) and (x_2, y_2) is $\sqrt{(x_2 - x_1)^2 + (y_2 - y_1)^2}$.

distributive law   For any real numbers a, b, and c, $a(b + c) = ab + ac$.

domain (relation)   The set of all possible first members of the ordered pairs that comprise a relation.

domain (variable)   The set of all possible replacements for a variable.

element   A member of a set or an entry in a matrix.

ellipse   The set of all points P in a plane such that the sum of the distances from P to two fixed points, called *foci*, is constant.

equilateral hyperbola   The hyperbola $xy = k$ $(k \neq 0)$ whose asymptotes are the coordinate axes. Also called a *rectangular hyperbola*.

equivalent equations Equations that have the same solution set. Thus, $2x = 6$ and $x = 3$ are equivalent equations.

even function Function f is even if $f(-x) = f(x)$ for all x in the domain of f.

exponent In x^n, the number n is the exponent and is the number of times the base x is used as a factor in a product. Thus, $x^3 = x \cdot x \cdot x$.

exponential equation An equation in which the variable appears in an exponent.

exponential function A function of the form $y = b^x$ where b is a positive constant that is not equal to 1.

factor A number or variable that is being multiplied in a product.

factor theorem The binomial $x - r$ is a factor of the polynomial $P(x)$ if $P(r) = 0$.

factoring The process by which a number or polynomial is written as the product of two or more terms.

factoring completely Factoring a number or polynomial into its prime factors.

factorial n Denoted by $n!$ and defined for any positive integer n as the product of consecutive integers from n to 1. Thus, $5! = 5 \cdot 4 \cdot 3 \cdot 2 \cdot 1 = 120$. By definition, $0! = 1$.

field A set of numbers and two operations that are defined on the set such that there exist identity and inverse elements for each operation and the closure, commutative, associative, and distributive properties hold. The set of real numbers forms a field with respect to the operations of addition and multiplication.

focus (foci) A fixed point or points that help to determine the shape of an ellipse, hyperbola, or parabola.

FOIL The rule for multiplying two binomials horizontally.

function A relation in which no two ordered pairs have the same first member and different second members.

fundamental theorem of algebra A polynomial equation has at least one root, real or nonreal. An nth-degree polynomial equation has exactly n roots, provided each root is counted as many times as it occurs. The roots may be real or imaginary. If a polynomial equation with real coefficients has imaginary roots, then they occur in conjugate pairs.

geometric sequence A sequence of numbers or terms in which each term after the first is obtained by multiplying the term that precedes it by the same constant, called the *common ratio*. The nth term of a geometric sequence is $a_n = a_1 r^{n-1}$, where a_1 is the first term and r is the common ratio.

geometric series The indicated sum of the terms of a geometric sequence. The sum of the first n terms of a geometric sequence is

$$S_n = \frac{a_1(1 - r^n)}{1 - r}.$$

greatest common factor (GCF) The GCF of two or more monomials is the monomial with the greatest coefficient and the variable factors of the greatest degree that are common to all the given monomials. The GCF of $8a^2b$ and $20ab^2$ is $4ab$.

greatest integer function The function $f(x) = [x]$ whose function value is the largest integer that is less than or equal to x.

horizontal line test If no horizontal line intersects a graph in more than one point, the graph represents a one-to-one function.

hyperbola The set of all points P in a plane such that the absolute value of the difference of the distances from P to two fixed points, called *foci*, is constant.

identity An equation that is true for all possible replacements of the variable.

imaginary unit The number denoted by i where $i = \sqrt{-1}$.

imaginary number A number of the form bi where b is a nonzero real number and i is the imaginary unit.

inconsistent system A system of linear equations that has no solution.

independent variable For a function of the form $y = f(x)$, x is the independent variable.

index The number k in the radical expression $\sqrt[k]{x}$ is the root of x to be taken. In a square root radical the index is omitted and is understood to be 2.

inequality A sentence that expresses an inequality relation by using a symbol such as $<$ (is less than), $\leq$ (is less than or equal to), $>$ (is greater than), $\geq$ (is greater than or equal to), or $\neq$ (is unequal to).

infinite sequence A sequence whose domain (set of subscripts) is the set of natural numbers.

infinite series The indicated sum of the terms of an infinite sequence.

integer A number from the set $\{\ldots, -3, -2, -1, 0, 1, 2, 3, \ldots\}$.

intermediate value theorem If $P(x)$ is a polynomial with real coefficients and $P(a)$ and $P(b)$ have opposite signs, then a real root of $P(x) = 0$ lies between a and b.

inverse of a function The relation obtained by interchanging the first and second members of each ordered pair of the function. The result may or may not be a function.

inverse of a matrix The multiplicative inverse of a square matrix $\mathbf{A}$, if it exists, is the matrix $\mathbf{A}^{-1}$ such that $\mathbf{A}\mathbf{A}^{-1} = \mathbf{A}^{-1}\mathbf{A} = \mathbf{I}$, where $\mathbf{I}$ is the identity matrix.

inverse variation If y varies inversely as x, then $xy = k$ $\left(\text{or } y = \dfrac{k}{x}\right)$ where k is a nonzero number called the *constant of variation*.

irrational number A number that cannot be expressed as the quotient of two integers.

leading coefficient For a polynomial in one variable, the number that multiplies the variable with the greatest exponent.

linear equation An equation in which the greatest exponent of any variable is 1.

logarithmic function The inverse of the exponential function.

logarithm of x An exponent that represents the power to which a given base must be raised to produce a positive number x.

major axis The longer of the two axes of symmetry of an ellipse whose endpoints, called *vertices*, are points on the ellipse. The length of the major axis is denoted by $2a$.

mantissa The decimal part of a common logarithm.

mapping A pairing of each member of a set with exactly one member of another set.

mathematical induction A special method of proof used to prove or disprove formula type of statements that are expressed in terms of a positive integer n. See **principle of mathematical induction**.

matrix A rectangular array of numbers or terms called *elements*. An $m \times n$ matrix has m rows and n columns. The element in the ith row and jth column of matrix $\mathbf{A}$ is named a_{ij}.

matrix addition If $\mathbf{A}$ and $\mathbf{B}$ are $m \times n$ matrices, then $\mathbf{A} + \mathbf{B}$ is an $m \times n$ matrix in which each element is the sum of the elements in the corresponding positions of $\mathbf{A}$ and $\mathbf{B}$.

matrix multiplication If $\mathbf{A}$ is an $m \times n$ matrix and $\mathbf{B}$ is an $n \times p$ matrix, then $\mathbf{A}\mathbf{B}$ is an $m \times p$ matrix whose entry in the ith row and jth column is the sum of the products obtained by multiplying the elements in the ith row of matrix **A** by the corresponding elements in the jth column of matrix **B**.

midpoint formula The midpoint of the line segment whose endpoints are (x_1, y_1) and (x_2, y_2) is

$$\left(\frac{x_1 + x_2}{2}, \frac{y_1 + y_2}{2} \right).$$

minor axis The axis of symmetry of an ellipse that is perpendicular to the major axis at the center and whose endpoints are on the ellipse. The length of the minor axis is denoted by $2b$.

monomial A number, variable, or the product of a number and a variable.

multiplicative identity In the set of real numbers, 1 is the multiplicative identity element since the product of any real number and 1 is that number.

multiplicative identity matrix For any square matrix **A**, the square matrix **I** that has 1 along its main diagonal and 0 everywhere else since $\mathbf{AI} = \mathbf{A}$.

multiplicative inverse In the set of real numbers, the multiplicative inverse of a nonzero number is the reciprocal of that number since their product is 1.

natural exponential function The function $f(x) = e^x$ where e is the special irrational number approximately equal to $2.718\ldots$.

natural logarithmic function The inverse of the natural exponential function, denoted by $\ln x$.

odd function Function f is odd if $f(-x) = -f(x)$ for all x in the domain of f.

one-to-one function A function in which no two ordered pairs of the form (x, y) have the same value of y and different values of x.

ordered pair Two numbers that are written in a definite order.

ordinate The y-coordinate of a point in the coordinate plane.

origin The zero point on a number line.

parabola The set of all points in a plane whose distances to a fixed point, called the *focus*, are the same as their distances to a given line, called the *directrix*. The vertex of a parabola is the point at which its axis of symmetry intersects the curve. The directed distance between the vertex and the focus is denoted by p.

Pascal's triangle A triangular array of numbers representing the values of $_nC_r$. The row numbers of the triangle correspond to successive values of n starting with $n = 0$ where $r = 0$ refers to the first number on each row. Thus, the value of $_4C_2$ is the third number in row 4 of Pascal's triangle.

perfect square A rational number whose square root is rational.

permutation An ordered arrangement of objects.

point-slope form of a linear equation The equation of a nonvertical line whose slope is m and that contains the point (x_1, y_1) is $y - y_1 = m(x - x_1)$.

polynomial A monomial or the sum or difference of two or more monomials.

polynomial function A function P such that $P(x)$ equals a polynomial in variable x. When the powers of the terms of the polynomial are written in descending order, the polynomial function is in standard form. The leading coefficient is the nonzero number that multiplies the variable with the greatest exponent.

power   A number or variable written with an exponent.

prime factorization   The factorization of a polynomial into factors each of which is divisible only by itself and 1.

principle of mathematical induction   To prove a formula is true for all positive integer values of n, complete these two steps:

(1) Show the formula is valid for the smallest possible value of n, usually $n = 1$.

(2) Assume the formula is true for some arbitrary positive integer, say $n = k$, and then prove the formula is true for the next consecutive integer, $n = k + 1$. This proves that, since the formula is true for $n = 1$, it is true also for the next integer, $n = 2$. Since the formula is true for $n = 2$, it is true also for the next integer, $n = 3$, and so forth.

proportion   An equation that states that two ratios are equal. In the proportion $\dfrac{a}{b} = \dfrac{c}{d}$, the product of the means equals the product of the extremes. Thus, $b \cdot c = a \cdot d$.

quadrant   One of four rectangular regions into which the coordinate plane is divided.

quadratic equation   An equation that has the form $ax^2 + bx + c = 0$, provided $a \neq 0$.

quadratic formula   If $ax^2 + bx + c = 0$, then

$$x = \frac{-b \pm \sqrt{b^2 - 4ac}}{2a} \quad (a \neq 0).$$

quadratic polynomial   A polynomial whose degree is 2.

radical sign   The symbol $\sqrt{} \cdot$

radical equation   An equation in which the variable appears underneath a radical sign.

radicand   The expression that appears underneath a radical sign.

range   The set of all possible second members of the ordered pairs that comprise a relation.

ratio   A comparison of two numbers by division. The ratio of a to b is the fraction $\dfrac{a}{b}$, provided $b \neq 0$.

rational function   The quotient of two polynomial functions.

rational number   A number that can be written in the form $\dfrac{a}{b}$ where a and b are integers and $b \neq 0$.

rational root theorem   If $\dfrac{p}{q}$ is a rational root in lowest terms of a polynomial equation with integer coefficients, then p is a factor of the constant term of the polynomial and q is a factor of the leading coefficient of the polynomial.

real number   A number that is a member of the set that consists of all rational and irrational numbers.

relation   A set of ordered pairs.

remainder theorem   The remainder obtained by dividing a polynomial $P(x)$ by $x - r$ is $P(r)$.

replacement set   The set of values that a variable may have.

root   A number that makes an equation a true statement.

scalar multiplication   Multiplication of a matrix by a constant.

sequence   An ordered list of numbers or terms.

sequence function   A function whose range is the set of unordered terms of a sequence and whose domain is the set of their position numbers.

Σ (sigma)   The Greek letter Σ, which indicates a summation of terms.

slope formula   The slope of a nonvertical line that contains the points (x_1, y_1) and (x_2, y_2) is $\dfrac{y_2 - y_1}{x_2 - x_1}$.

solution set   The collection of all values from the replacement set of a variable that make an equation a true statement. Each of these values is a root of the equation.

square matrix   A matrix that has the same number of rows as columns.

square root   The square root of a nonnegative number n is one of two identical numbers whose product is n. The symbol $\sqrt{x}$ denotes the positive square root x. Thus $\sqrt{9} = 3$ since $3 \times 3 = 9$.

synthetic division   A method of dividing a polynomial $P(x)$ by a binomial of the form $x - r$ that uses the detached coefficients of the powers of x.

system of equations   A set of equations whose solution, if it exists, is the set of values that make each of the equations true at the same time.

theorem   A generalization in mathematics that can be proved.

triangular form of a matrix   A matrix in which all the elements below the main diagonal are 0.

trinomial   A polynomial with three unlike terms.

transverse axis   An axis of symmetry of a hyperbola whose midpoint is the center of the hyperbola and whose endpoints are points on the hyperbola called *vertices*. The length of the transverse axis is denoted by $2a$.

variation in sign   Occurs when two consecutive coefficients of a polynomial in standard form have different signs.

vertex   The point at which the axis of symmetry intersects a parabola.

vertical line test   If no vertical line intersects a graph in more than one point, the graph represents a function.

vertices (ellipse)   The points on the ellipse that are the endpoints of the major axis.

vertices (hyperbola)   The points on the hyperbola that are the endpoints of the transverse axis.

zero of $P(x)$ Any value of x such that $P(x) = 0$.

zero product rule   If the product of two expressions is 0, then at least one of them must be 0.

INDEX

BARRON'S COLLEGE REVIEW SERIES
Excel in Your Course

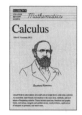

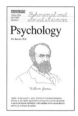

Each title in *Barron's College Review Series* offers you an overview of a college-level course, and makes a fine supplement to your main textbook. You'll find topic summaries, lists of key terms, bibliographies, review questions, and more.

Algebra
S. L. McCune, E. D. McCune
and J. R. Turner
ISBN 9746-7 $16.99, Can.$24.50

Calculus
E. C. Gootman
ISBN 9819-6 $16.95, Can.$24.50

Chemistry
N. Jespersen
ISBN 9503-0 $16.95, Can.$24.50

Organic Chemistry
J. W. Suggs, Ph.D.
ISBN 1925-7 $16.95, Can.$23.95

Psychology
D. Baucum
ISBN 0674-0* $16.95, Can.$24.50

Statistics
Martin Sternstein, Ph.D.
ISBN 9311-9 $16.95, Can.$24.50

United States History
To 1877
N. Klose and R. F. Jones
ISBN 1834-6 $14.95, Can.$21.00

United States History
Since 1865
N. Klose and C. Lader
ISBN 1437-9* $18.99, Can.$27.50

Books may be purchased at your bookstore, or by mail from Barron's. Enclose check or money order for the total amount plus sales tax where applicable and 18% for postage and handling (minimum charge $5.95). New York, New Jersey, Michigan, and California residents add sales tax. Prices subject to change without notice.

Barron's Educational Series, Inc.
250 Wireless Blvd. • Hauppauge, NY 11788
In Canada: Georgetown Book Warehouse
34 Armstrong Ave., Georgetown, Ont. L7G 4R9
Visit our website at: www.barronseduc.com

ISBN Prefix: 0-8120, except where
followed by *, * = 0-7641 prefix.

$ = U.S. Dollars   Can$ = Canadian Dollars

R 6/05 (#58)